# Competing for the future

Everybody knows that digital technology has revolutionized our economy and our lifestyles. But how many of us really understand the drivers behind the technology – the significance of going digital; the miniaturization of electronic devices; the role of venture capital in financing the revolution; the importance of research and development? How many of us understand what it takes to make money from innovative technologies? Should we worry about manufacturing going offshore? What is the role of India and China in the digital economy? Drawing on a lifetime's experience in the industry, as an engineer, a senior manager, and as a partner in a global venture capital firm, Henry Kressel offers an expert personalized answer to all these questions. He explains how the technology works, why it matters, how it is financed, and what the key lessons are for public policy.

HENRY KRESSEL is a Managing Director of Warburg Pincus, LLC. He began his career at RCA Laboratories where he pioneered the first practical semiconductor lasers. He was the founding president of the IEEE Laser and Electro-Optics Society (LEOS) and co-founded the IEEE/OSA *Journal of Lightwave Technology*. He is the recipient of many awards and honors, a fellow of the American Physical Society and of the IEEE, and the holder of thirty-one issued US patents for electronic and optoelectronics devices.

THOMAS V. LENTO is founder and President of Intercomm, Inc., a corporate communications consultancy.

# Competing for the future

## How digital innovations are changing the world

HENRY KRESSEL
WITH
THOMAS V. LENTO

CAMBRIDGE
UNIVERSITY PRESS

CAMBRIDGE UNIVERSITY PRESS
Cambridge, New York, Melbourne, Madrid, Cape Town, Singapore, São Paulo

Cambridge University Press
The Edinburgh Building, Cambridge CB2 8RU, UK

Published in the United States of America by Cambridge University Press, New York

www.cambridge.org
Information on this title: www.cambridge.org/9780521862905

First published 2007

Printed in the United States of America

*A catalogue record for this publication is available from the British Library*

ISBN 978-0-521-86290-5 hardback

*For Bertha*

# Contents

# Figures

# Tables

# Acknowledgements

The idea for this book has its origins in a visit to the University of Cambridge in 2005, hosted by Professor Ian Leslie, Pro-Vice-Chancellor (Research). I am very grateful for the insights I gained from him regarding the technology programs at the university. I met some of the faculty and students of the Cavendish Laboratory, a place where so much of the world's basic scientific work was accomplished. I also visited a number of entrepreneurial companies founded to commercialize the innovations that grew out of university research. I am indebted to Lord Broers, former vice-chancellor of the university, and to my partner Dr. William H. Janeway, Chairman of Cambridge in America for making the appropriate introductions to facilitate my visit and the many valuable discussions.

These visits called to mind the long chain, bridging the centuries, that links fundamental scientific discovery to technological advances. This process has accelerated dramatically in the twentieth century, allowing digital technologies to transform the modern world in a remarkably short time. Why and how this has happened, and the consequences of the transformation, are among the topics that I have explored in this book.

In deciding to write this book, I have been fortunate in working with Thomas V. Lento, without whom it would not have been completed. His collaboration was critical in shaping its contents.

This book also reflects my experience over the years from collaborations in venture investments with my partners at Warburg Pincus: Lionel I. Pincus, John L. Vogelstein, Dr. William H. Janeway, Joseph P. Landy, Charles R. Kaye, Jeffrey A. Harris, Dr. Harold Brown, Bilge Ogut, Beau Vrolyk, Cary J. Davis, James Neary, Patrick T. Hackett, Henry B. Schacht, Steven G. Schneider, Stewart Gross, Robert Hillas, Dr. Nancy Martin, Julie Johnson Staples, Andrew Gaspar, Christopher W. Brody, and Frank M. Brochin.

My knowledge of China was built up over the years in collaboration with the Warburg Pincus team in Hong Kong – Chang Q. Sun, Jeff Leng and Julian Cheng.

Dr. Peter D. Scovell and Dr. Stanley Raatz greatly enhanced the accuracy of the material through their perceptive reading of the manuscript, and Chris Harrison, my editor at Cambridge University Press, provided essential guidance for refining its presentation.

In my early technical career I had the good fortune to work with many talented collaborators at RCA. These included Dr. Adolph Blicher, Herbert Nelson, Benjamin Jacoby, Dr. Jim J. Tietjen, Dr. Harry F. Lockwood, Dr. Michael Ettenberg, Frank Hawrylo, and Ivan Ladany. More recently, I have greatly benefited from my work with SRI International headed by Dr. Curt Carlson and Sarnoff Corporation staff, in particular Dr. Satyam Cherukuri.

Finally, I am very grateful to Tina Nuss for her extraordinary work in preparing and proofreading the manuscript.

# Introduction – Competing for the future: How digital innovations are changing the world

D AVID Sarnoff, long the chairman of RCA and a pioneer in the electronics industry, summarized his career in these words: "I hitched my wagon to the electron rather than the proverbial star."[1] The world has followed suit. The past sixty years have witnessed the most rapid transformation of human activity in history, with digital electronic technology as the driving force.

Nothing has been left untouched. The way people communicate, live, work, travel, and consume products and services have all changed forever. The digital revolution has spurred the rapid expansion of economic activity across the face of the planet.

In this book I will explore the unprecedented outburst of electronic innovation that created the digital revolution. Based on this example, I will examine how innovation works, what it has achieved, and what forms we can expect it to assume in the near future.

Since innovation does not happen in a vacuum, I will also explore the political and economic factors that can accelerate or impede changes in the industrial landscape. One of these is globalization, which creates neither a level playing field nor a truly "flat world." Governments everywhere are focused on industrializing as quickly as possible in the face of growing competition. As a result, attempts to gain national competitive advantage by building artificial walls are not going away.

## Defining innovation

Before outlining the issues to be covered, we must clarify what constitutes an "innovation" in the first place. The blue laser diode that

---

[1] The *New York Times*, April 4, 1958.

enables the next generation of optical storage is obviously an innovation. But many would say an improved connector for circuit boards
qualifies too. To avoid confusion, we need a definition that focuses the
term on the kind of transformational technology that we're exploring.
Here is a good starting point.

Innovation is an historic and irreversible change in the way of doing things . . .
This covers not only techniques, but also the introduction of new commodities, new forms of organization, and the opening of new markets.[2]

In this book, then, innovation refers to a technical development or
invention that contributes toward creating new industries, new products and processes, and new approaches to solving complex industrial
problems. Innovation includes commercializing the technology to
move it into the market and disseminate it throughout the world.

Needless to say, some technical developments or inventions are more
fundamental and far-reaching than others. The most potent force for
change is a revolutionary innovation. This is a basic invention that
ultimately replaces a technology once central to industry.

A revolutionary innovation is the tsunami of the industrial world. Its
power spreads out from a central technology base, overcoming all
resistance in its path through successive industrial innovations, until
its effects are felt in the farthest reaches of the world economy.

Revolutionary innovations are followed and supported by innovations that are valuable in and of themselves, but are more evolutionary
in nature. Evolutionary innovations help implement the fundamental
discoveries and extend their reach.

### Building the digital world

To understand how digital electronics conform to this pattern, we must
first look at how the technology was developed. Then we will examine
how it has affected industrial development around the world.

It promises to be a fascinating journey. The following overview can
serve as a map of the terrain, to guide us through the issues we will
encounter. To get started, let's orient ourselves by looking at where
electronic technology came from in the first place.

---

[2] J. A. Schumpeter, *Readings in business cycle theory* (Philadelphia: The Blackiston
Company, 1944), p. 7.

Modern electronics began with the invention of the vacuum tube (called a "valve" in the UK) in the early twentieth century. This device initiated the "analog" electronic age, from radio through telephony, television, radar, and the very beginnings of computing.

Today vacuum tubes have all but disappeared. They were replaced by two of the most revolutionary device innovations in history, the transistor and the semiconductor laser diode. These landmark inventions proved to be the key components in the digital technologies that have transformed the world. Chapter 1 talks about them in more depth.

Significantly, both the transistor and the laser diode were largely conceived and commercialized in the United States. Following World War II, the US had become the epicenter for technological breakthroughs, the result of enormous R&D investments by major corporations over a span of many years. These companies also carried out the commercial implementation of their discoveries.

US technical leadership persisted through the end of the twentieth century. As we shall see, however, in the later years this was due more to financial and social structures than industrial and governmental support. In fact, it was the availability of private venture capital in the United States during the 1980s and 1990s that drove many revolutionary innovations to their present dominance.

Venture funding made possible the formation of hundreds of highly innovative companies that created new digital electronic products to pioneer new markets. They produced innovations in communications and computing, particularly microprocessors and software, that generated whole new spheres of activity.

Now, at the start of the twenty-first century, the world order has changed. Innovation and the capital to finance new technology businesses are more globally distributed. This has profound consequences for regional economic development.

## Innovation on a global scale

Today the talk is all about globalization, as if it were a new phenomenon. But globalization has been going on for thousands of years, with people migrating across lands and seas, carrying ideas and products with them. The real difference in the modern world is how much faster the process has become.

The invention of printing in the fifteenth century led to an enormous increase in the spread of ideas, but it took about 100 years to make meaningful changes to Western society. The introduction of the steam engine in the eighteenth century triggered its own chain of industrial transformations, but it wasn't until the middle of the nineteenth century that its full impact began to register on societies around the world.

By contrast, digital electronics have profoundly altered societies around the globe in little more than twenty years. This has happened despite the fact that the electronics revolution has involved countries that were not full participants in the earlier industrial upheavals.

Part of the reason is political. After the two world wars of the twentieth century, Europe and Asia underwent nearly twenty years of reconstruction. Meanwhile, US businesses were becoming the undisputed leaders in technological industries. As we will see, the largest corporations established central research laboratories in the expectation that fostering industrial innovations would help assure their futures.

By the 1970s, however, we see the beginning of a new wave of very rapid globalization. The industrialized countries most devastated by the war in Western Europe had recovered, as had Japan, and they began to build their industries using technologies that had been largely innovated in the United States. They represented new, well-capitalized international competitors in the world market.

That was a rude awakening for the United States. Suddenly, industry leaders found themselves faced by formidable challengers. Complacency quickly gave way to panic. Most noticeable was the challenge mounted by Japanese companies, which were competing not only on price but also on quality. Forecasts of the inevitable takeover of US industry by large Japanese companies produced widespread public fear.

The feared collapse of the US economy under the onslaught of European and Japanese competition did not happen. What did occur was the rise of a whole new class of formidable Asian competitors to further threaten the dominance of incumbents. Starting in the late 1970s, South Korea, Taiwan, Malaysia, and the Philippines became centers of low-cost manufacturing outsourced from the US.

No one dreamed at the time that China would emerge as an industrial colossus any time soon. Its economy was too restrictive and too badly managed. But emerge it did, starting in the 1980s, when its government decided to open the country to market forces, entrepreneurial capitalism, and, increasingly, foreign investment.

Nor did anyone predict that India would become a major source for the international production of software. But today Bangalore is booming with software firms.

The acceleration of industrial development in the leading Asian economies has been astonishing. They have accomplished in just a few years what once took many decades. How were they able to manage it?

I will propose that this shock wave of rapid development resulted from a crucial combination of factors not present in previous phases of industrialization. First and most important was the readily transferable nature of the technology being commercialized. Although the revolutionary innovations in digital electronics originated primarily in the US between the 1940s and 1980s, they allowed an extraordinary degree of global industrial mobility.

Other factors included new government policies in the industrializing nations, and the availability of risk capital to fund new business ventures. We will consider the effectiveness of both approaches in fostering industrial development.

## The new world order

Earlier waves of globalization saw industries move to new parts of the world. New centers of innovation arose. The movement of risk capital seeking the highest return financed industrial development in those new regions. We have witnessed the same train of events in our own time, but in a highly compressed time frame.

We have also seen the inevitable consequences of this process. Excessive investment overbuilds capacity, and a return to reality is accompanied by a destructive collapse. The boom and bust in telecommunications in the late 1990s, where trillions of dollars of value were wiped out in a few short years, is a classic example.

As the aftereffects of the tidal wave of digital electronics ripple in new directions, it is time to take stock of what has happened, why, and how. I trust that this book will provide a useful analysis.

More than that, I hope it will stimulate discussions not just about the future of technology, innovation, and industrialization, but about their global impact. One thing is certain: the future will be quite different from the linear extrapolations of the present.

In talking about the future, I am mindful of the spotty history of technological and market predictions. One of my favorite examples involves conflicting views on television. In 1939 David Sarnoff, the chairman of RCA, presided over the televised opening of the RCA Pavilion at the World's Fair. He said: "Now we add sight to sound. It is with a feeling of humbleness that I come to this moment of announcing the birth, in this country, of a new art so important in its implications that it is bound to affect all society. It is an art which shines like a torch in a troubled world."

Contrast his prophetic statement with the opinion expressed in a *New York Times* editorial at about the same time: "The problem with television is that people must sit and keep their eyes glued to the screen; the average American family hasn't time for it. Therefore the showmen are convinced that, for this reason, if no other, television will never be a serious competitor of [radio] broadcasting." Who would have guessed that the average household now spends over six hours a day watching television?[3]

---

[3] Quoted in the *New York Times*, August 28, 2005, Opinion, 12.

# PART I

# The technology – how electronic devices work – digital systems and software

# 1 | *Genesis: Inventing electronics for the digital world*

THIS book is about the creation and consequences of digital electronics.

Over the last half-century no other area of technical innovation has so drastically altered the way we live, work, and interact. Digital electronics give us instant access to a whole universe of information, and powerful tools to process it. They also equip us with an unprecedented ability to communicate instantly with people anywhere in the world.

I was privileged to contribute to these developments, first as a scientist engaged in creating new electronic devices and systems, and then as a venture capitalist involved in creating successful new businesses.

From these vantage points, I watched digital electronics grow from a technical specialization into a revolutionary force that generates new industries and transforms developing economies into global competitors, and does both seemingly overnight.

To sense the full extent of its impact, first visualize a contemporary scenario. A programming group in India is working on a project from the US. They are using computers built in China, with LCD displays made in Korea, driven by software from Silicon Valley and Germany. They communicate with colleagues in the US and other countries over the Internet, by e-mail and voice. It all seems quite normal today.

Now try to picture the same situation twenty-five years ago. You can't. There was little global sourcing of technical expertise then. High-tech manufacturing in developing Asian economies was just getting started. The major sources of equipment were Japan, the US, and Europe. Collaboration over the Internet was a science-fiction fantasy. Digital electronics had yet to transform our lives.

That's how fast and how thoroughly things have changed. If we're to track these developments and their implications, there is obviously a lot of ground to cover.

Where should we begin our journey? My choice is to start where I began my career, with solid-state devices, which were the genesis of the whole field of digital electronics.

Why start with these early innovations? Because so much of the progress in digital systems grew out of the advances in structures and materials embodied in these basic devices. An understanding of how they work gives you a better grasp of the evolution of digital technology. By knowing their physical characteristics you can also better appreciate the limitations that will sooner or later put the brakes on further progress.

The crucial factor of device miniaturization is a prime example. When I was working on electronic devices, miniaturization was a dream endeavor for materials scientists. But its significance extends far beyond the solution of a technical challenge.

The drive to miniaturize led to the concept of reducing the size and cost of digital systems. Miniaturization has been driving performance increases in semiconductor devices and magnetic disk storage for the past forty years. Marvelous advances in software, after all, need an ever-increasing number of transistors running at ever higher speeds to process their commands.

By constantly reducing device sizes, scientists have kept systems small and inexpensive while their power and complexity have increased. The fact that people can carry and afford to own sophisti-cated cell phones that fit in their pockets is a direct outcome of the effort to miniaturize devices.

Today, unfortunately, even technologists ignore the roots of digital technology. I have interviewed job applicants with backgrounds in computer science and digital systems for many years. More often than not, I ask them to explain to me how a transistor works.

What I usually get in response is a sage discourse on Moore's Law and how this inexorable law of nature will continue to drive up the value of digital systems for the foreseeable future. When I remark that Moore's Law is neither a law of nature nor inexorable, and press them again as to just how a transistor *really* works, I'm often answered by an embarrassed silence.

The fact is that everyone, engineers included, takes the basic devices for granted. They're looked at as immutable, like the rising and setting of the sun. Yet when it comes to continuing the progress of digital electronics, this is where the real energy lies. It's the power dissipation of devices that sets serious limits on future performance gains in digital systems. I take pains in this chapter and the next to explain why and how these limitations will be circumvented.

So we open our study with some material to help you understand the physical basis of digital electronics. At the end of the first two chapters you can take some satisfaction in knowing that you're better versed on the subject than some engineers I've met, and it's only taken you an hour to get there.

In this first chapter I have also included an introduction to the next wave of device technology, based not on semiconductors but on polymers, which are organic materials. No one dreamed that this would happen forty years ago when technologists worked only on crystalline semiconductors.

## An insider's perspective

I confess to some bias toward technical information. I started my career at RCA as a physicist fascinated by solid-state devices. This passion remained with me throughout the many projects I worked on, including these developments:

- the company's first production silicon transistor (1960);
- integrated circuits (1961);
- microwave devices for the Apollo mission radios (1961);
- practical semiconductor lasers (1967);
- solar cell energy converters (1974);
- fiber optic systems for military applications (1975);
- high-efficiency light sensors for fiber optic systems (1979).

It was my good fortune to be in the right place to make these contributions. I was at RCA at a time when that company prided itself on fostering electronic innovations and was willing to invest the money necessary to take them to market.

At RCA Laboratories, I was surrounded by remarkable researchers. In the 1960s and 1970s, my friends down the hall there developed the modern MOSFET (metal oxide semiconductor field-effect transistor), CMOS (complementary metal oxide semiconductor) integrated circuits, the solid-state imager, and the first liquid crystal displays. These inventions are at the heart of the digital devices and systems we use every day.

We knew that those were remarkable times. But none of us could have predicted where our work and that of others in our nascent fields would lead. Our industry conferences focused on technical developments, not on multi-billion dollar industries that would change the world's economic order. The business impact came much more slowly than the technical breakthroughs.

Looking back, I wonder at the miraculous way that so many revolutionary innovations, originally intended for applications that appeared obvious at the time, ended up coming together in totally unexpected ways to create the digital systems described in Chapter 2.

If there is an equivalent to Adam Smith's "invisible hand" of the market that somehow guides the integration of all these technologies, it keeps its own counsel. The development of these technologies was by no means a coordinated effort. The innovations originated in various organizations, each of which made its own investment decisions.

Many of those decisions were proved wrong, and the companies failed to reap any commercial rewards for their inventions. Of course they didn't have our historical perspective on technology markets to help in their decision making. They were sizing up immediate opportunities and taking action on the basis of incomplete information.

## *The case for vision*

It is fashionable to believe that revolutionary innovations inevitably lead to world-changing events, and that visionaries can see beyond present conditions to predict which inventions will be successful. We will have occasion in Chapter 3 to discuss industry leaders who successfully staked large bets on the potential of revolutionary innovations to open new markets. But such visionary leaders are rare.

All too often linear thinking, the projection of current knowledge into the future, is confused with genuine vision. I remember looking at a book from the 1920s on how commercial aviation would look in fifty years. The author foresaw giant planes with indoor lounges and swimming pools. They would have dozens of engines that would be serviced by technicians walking on the wings (in flight) to keep the oil level up.

In effect he was describing a big passenger ship with wings. The author had no other point of reference for his predictions except ocean liners and planes with unreliable small engines. This is classic linear thinking.

We all have stories to tell on this score. In my case, in the early 1970s, I showed our RCA Laboratory fiber optic communications link to an expert on communications. He commented that there was going to be a very small market for such systems.

He was sure that one or two optical links at 1 Gb/s between San Francisco or Los Angeles and New York would be enough to serve all the communications needs of the US. Such systems would need just one laser at each end. Thus semiconductor lasers had no commercial future except as aiming devices for military systems.

Fortunately, this individual was not in a position to influence the direction of RCA's research and I continued on my way. However, all too often there are instances where such thinking at high management levels in big companies kills programs.

For example, AT&T's management decided against managing the predecessor to the Internet (the ARPANET) in the 1970s, believing that switched-packet networks could never compete with their existing telephony switched network.[1]

The reason for what now seems like misguided thinking is not hard to find. While the Department of Defense had a vital interest in building a data network capable of withstanding hostile attacks, AT&T's telephony network had no such requirement and was in fact very reliable. Hence, why bother with a risky new technology that would disturb things?

Of course, it helps to be a monopoly, and the incipient Internet posed no threat to the company's existing business. As a result, AT&T's Bell Labs was late to begin research on Internet Protocol systems, leaving others to take market leadership. (Bell Labs did, however, contribute enormously to device technology and computer technology with the invention of the UNIX operating system for high-performance computing.)

The real visionaries in the early days were to be found in US defense organizations. They were funding truly innovative electronics research for practical applications in national defense.

In fact, military requirements drove the origin and implementations of many of the innovations that I talk about in this book. The Department of Defense (DoD) funded work on semiconductor lasers, integrated circuits, liquid crystal displays, imaging devices, fiber optic communications, advanced software, and computer systems, to name just some of the important projects.

---

[1] J. Naughton, *A brief history of the future: The origins of the Internet* (London: Weidenfeld and Nicolson, 1999), pp. 114–117.

This research was funded at corporate laboratories (at RCA Labs we were major recipients), at universities and their affiliated labs (such as the MIT Lincoln Laboratories), and at start-up companies with innovative ideas.

Were the funders at the DoD visionaries? Yes, in the sense that they recognized that only *revolutionary innovations* could address pressing problems not amenable to the solutions of linear thinking. They also had the good sense to risk funding a wide variety of research projects. Many failed to come up with anything useful. But the spectacular successes more than made up for the failures.

## Innovation as a social force

The rest of this chapter will describe the basic concepts and significance of transistors, integrated circuits, memories, semiconductor lasers, fiber optics, and imagers and displays.

These devices did not leap into the world fully formed, and some took decades to find their true role in fomenting the digital revolution. Going from concept to practical products is a long stretch.

Their migration through intervening applications is not important in the context of this book. My major purpose is to communicate the impact that these innovations have had on broader economic and social developments. Also, since the book is not intended as a history of science and technology, I have done my best to keep its focus on the practical realizations of innovations.

We now turn our attention to the origins of our age.

## Where it all started

Fifty years ago the most complex electronic product in the average American home was a black-and-white television. It usually received only three or four channels. Coverage of breaking news was limited to a studio announcer reading reports to viewers at 6:00 p.m. If the story was big enough, the announcer would patch in a telephone conversation with a reporter at the scene.

Compare that with the media-rich, information-intensive environment of the twenty-first century. Is there a war, an earthquake, an Olympic triumph, an astronaut arriving at the space station? Hundreds of sources can put us in the middle of the action with live audio and video, in full color and high definition. We can watch the story break on 24-hour cable and satellite news channels, or on Web sites we access through our personal computers.

Miracles like this are now so commonplace that we take them for granted. In the 1950s, calling someone in the next state was "long distance." In 2006, it's easier and cheaper to phone Paris, Tokyo, or Islamabad. We can do it while walking down the street. As recently as the mid-1980s, computers were too arcane and expensive for ordinary mortals. Today we carry around laptop PCs with millions more circuit devices and vastly greater power than the behemoth corporate number crunchers of 1980.

It's a story of power made small, solid-state, and digital. By packing millions of circuits onto a sliver of silicon, innovators have made it possible for miniaturized devices to handle the high-speed information processing that drives today's digital functions.

The digital revolution has altered more than the size and power of our electronic tools. It has changed our way of life. New industries have emerged to exploit its power. Thanks to electronic communications, countries once considered backward or isolated have been able to become significant players in global commerce. A world community has emerged based on the Internet.

Social structures, government policies, business efficiency – all have been transformed by the computing and communications capabilities made possible by innovations in electronics.

## Building blocks of modern electronics

This transformation is all the more astonishing in that it sprang from a handful of electronic devices invented between 1947 and 1968. They include:

- The transistor, a semiconductor device to switch and amplify currents.
- Devices that store and retrieve large amounts of digital data, using either magnetic or solid-state technology.
- Laser light sources, light sensors, and glass fibers for optical signal transmission, which together enable digital fiber optic communications.
- Semiconductor devices to capture images in electronic form.
- Solid-state displays for images and data. The newest displays show the growing importance of polymer materials in electronic devices.

*Table 1.1 Key device innovations and their commercial introduction*

| Innovation | Year | Company | Notes |
|---|---|---|---|
| Transistor | Early 1950s | AT&T (Western Electric) | a |
| Magnetic Disk Storage | 1960 | IBM | |
| Integrated Circuits (ICs) | 1960 | Fairchild Semiconductor | b |
| CMOS ICs | 1969 | RCA | |
| Heterojunction Laser Diode | 1969 | RCA | c, d |
| Fiber Optics | Early 1970s | Corning | e |
| Liquid Crystal Displays | 1968 | RCA | f |
| CCD Imager | 1974 | Fairchild Semiconductor | g |

a – AT&T produced transistors for its own equipment. RCA, Sylvania, Philco, Texas
   Instruments (and others) introduced commercial germanium transistors in the 1950s.
b – Independent development at Texas Instruments.
c – Early laser diode research at General Electric, Lincoln Labs, and IBM Labs.
d – Heterojunction laser research independently conducted at RCA Labs, Bell Labs,
   and A. F. Ioffe Institute.
e – Fiber optic research also conducted at Bell Labs and Standard Telecommunications
   Laboratory, UK.
f – RCA introduced a commercial small alpha-numeric display for watches and
   instruments.
g – Basic CCD imager concept originated at Bell Labs. W. Boyle and G. Smith, US
   Patent 3,792,322, "Buried channel charge coupled devices," (February 12, 1974).
   Seminal research was also conducted at Philips.

These revolutionary innovations quickly swept away existing approaches, replacing them with technology that offered vast potential for further development.

Table 1.1 shows where the most important devices were developed in their ultimate practical form and taken to market. You will note that most grew out of efforts at large American corporations, whose staff conceived the technologies and developed manufacturing processes to bring them to market.

Table 1.1 does not attempt to list all of the individual inventors for these devices because our focus is on the socioeconomic impact of innovation, not the history of science. We will recognize those individual achievements in the text as appropriate, and when attribution is clear.

A word is needed about the difficulty of assigning individual credit for innovations. Students of the history of science and technology are well aware that practically no significant invention springs in isolation from the mind of a single inventor. Important innovations arise when creative and *prepared* minds understand and synthesize early results and meld them with their own inspirations into a radical new result.

For this reason more than one person can come up with some form of the same invention at the same time. Alexander Graham Bell and Elisha Gray both invented the telephone, but Bell got to the patent office first. Likewise, some of the inventions listed in Table 1.1 have been attributed to more than one person. In these cases, history will sort out the claims at some future time.

## From tubes to transistors

The true start of the electronic age can be dated from the invention of the electron *vacuum tube* (also called a *valve*) in the early years of the twentieth century. Tube-based technologies dealt with information almost exclusively in *continuous* (analog) form, mirroring how physical phenomena such as light and sound occur in nature.

This is very different from the solid-state electronic technologies which have transformed the world over the past fifty years. These rely primarily on the processing of information in *binary* (digital) form. This means they perform all of their computations using a numbering system with only two values, either a zero or a one.[2]

Since there were no alternatives to vacuum tubes until the 1940s, tubes and analog information processing provided the only practical approach to electronics for nearly a half-century. During that time, innovations in tube technology produced increasingly sophisticated devices to amplify and switch electronic signals, display images, process voice, and emit and receive radio signals. These devices were the

---

[2] A comparison of the analog method of sound recording, invented by Edison, with the digital approach will make the difference clear. In analog recording, audio waveforms are physically engraved on a phonograph disc as an undulating spiral groove. On a digital compact disk (CD) recording, however, the sound is represented by a "sample" of the signal, a series of ones and zeros which the CD player processes to reconstruct the original waveforms for playback. It might seem preferable (more "faithful") to use the simple, direct approach of keeping waveforms in the analog domain from recording through playback. But this exposes the signal to noise, interference, and gradual degradation. Converting it to digital bits preserves its integrity, with the added benefit of making it easy to enhance through computer processing.

enablers of all electronic systems: radio, television, telephones, telegraph, wireless communications, and radar.

Tubes were even used as digital switching devices on a large scale in
the early digital computers of the 1940s and 1950s. There were as
many as 60,000 tubes in one computer. But although digital processing
was technically possible with tubes, it was highly impractical. Early
tube-based computers were nightmares of size and heat generation, and
had very poor reliability.

By the 1950s, vacuum tube technology had reached its practical
limits. Tubes suffered from a fundamental handicap: they were single-
function devices incapable of efficient integration. Electronic products
such as radios required a large number of discrete tubes, wired with
resistors, inductors, and capacitors, to perform all of their functions,
making them bulky and power-hungry.

Tubes had other handicaps. They could not be miniaturized past a
certain point because of their principle of operation, which required an
electron "cloud" surrounding a cathode, anode, and other structures in
a vacuum. In addition, the inevitable deterioration of the internal
materials used to generate electrons in vacuum tubes limited their
reliability.

As a result, tube-based systems were fated to be supplanted by newer
technologies in the second half of the twentieth century. Electron tubes
have not completely disappeared. They have retained niche applications in microwave systems and in very expensive home audio systems
because of their unique properties, and the cathode-ray tubes (CRTs),
long used as display devices for TVs and computers, are vacuum tubes.
However, tubes are becoming much less important even in those areas.

## The first solid-state devices

The technology that was to re-energize electronic development
appeared in a primitive form: the *transistor*, invented by AT&T's Bell
Laboratories scientists in 1947.[3]

Their invention consisted of a tiny slab of germanium with two metal
point contacts making up the emitter and collector parts of the device.
The contacts corresponded to the cathode and anode of the vacuum

---

[3] J. Bardeen and W. H. Brattain, "The transistor, a semiconductor triode," *Physical
Review* 74 (1948), 230–231.

1.1. The evolution of vacuum tubes by miniaturization. The third tube from the left is the RCA Nuvistor. To the right of the Nuvistor is the first RCA mass-produced silicon transistor (2N2102) (David Sarnoff Library, ref. 4).

tube. But it was not just its small size that made the transistor so groundbreaking.

The transistor was the first device to replicate a vacuum tube's current amplification and switching functions using only solid-state (semiconductor) elements. It made the vacuum tube and its electron cloud unnecessary. From a technical standpoint the transistor was revolutionary because it showed that it was possible to manage electron flow within tiny bits of solid material. The achievement earned its inventors a Nobel Prize.

The electron tube industry, including RCA, General Electric, and Sylvania immediately recognized that a formidable new competitor was on the scene. It appeared just as the miniaturization of vacuum tubes had reached its peak.

For example, a new tube introduced by RCA, the Nuvistor, was remarkable for its small size and ruggedness, as Figure 1.1 shows.[4] Note that the transistor, packaged in its metal can, is almost as large as the Nuvistor, although the actual solid-state device inside the can is tiny.

[4] Photo courtesy David Sarnoff Library.

This transistor, the 2N2102, was RCA's first mass-produced silicon transistor (1961). Its development was my first project when I joined RCA.[5] It is still available from ST Microelectronics.

Even though the future of transistors was unknown, their potential as tube replacements was clear. As a result, all the major electron tube manufacturers started semiconductor business units in the 1950s. RCA, the leader in tubes, demonstrated experimental high-frequency transistors for use in radios and other products as early as 1952, and introduced its first commercial germanium transistor in 1957.

## Material for growth

Access to the new technology was easy and cheap. AT&T, which was not in the component business (except for its own use), licensed its transistor technology to anyone for only $25,000 as prepayment against modest royalties.

By 1958, the germanium transistor was widely available. Small portable radios based on the transistor were selling well. Computers were being designed with germanium transistors. The solid-state industry was born, though the full commercial impact of the device took years to be felt.

In the meantime Bell Laboratories had started work on silicon, the semiconductor material that was to prove dominant in the long run. This element was harder to master as part of the transistor production process, but it had the known advantage of withstanding higher operating temperatures than germanium.

The first generally useful silicon transistors appeared commercially in the late 1950s. Though it took time for them to displace germanium due to their initially high cost, falling prices and operational superiority eventually made them the devices of choice.

The development of silicon proved fortunate for the future of digital electronics. Although it received little attention in the 1950s, there was a crucial area in which silicon had a huge advantage over germanium: the fact that one could oxidize its surface (in a hot oxygen atmosphere)

---

[5] The 2N2102 evolved from the device described in H. Kressel, H. S. Veloric, and A. Blicher, "Design considerations for double diffused silicon switching transistors," *RCA Review* 4 (1962), 587–616.

to produce a thin insulating film of silicon dioxide (glass) with remarkable interfacial properties.

A whole new device technology enabled by this property eventually culminated in integrated circuits. Modern microelectronics are built around the unique properties of this oxide/silicon interface. This led directly to the incredible feats of miniaturization that made the digitization of electronics practical.

The oxides of germanium, on the other hand, were not useful, another reason the element faded as an important semiconductor material.

When I arrived at the RCA Solid State Division in late 1959 as a transistor designer, the heroes of the organization were the people working on germanium transistors. I joined the organization with no transistor or device experience, having answered an ad for junior engineers to work on new silicon transistor products being planned in a department headed by Dr. Adolph Blicher, a talented Polish scientist.

My job was to get a silicon transistor into production within one year. Although the Fairchild Semiconductor Corporation had just introduced some very expensive silicon transistors, the opinion of the local experts was that silicon was just too difficult to manufacture, and that germanium transistors would dominate the industry forever. The general view was that the new guy was wasting his time. They were right about the difficulty of mastering the technology, as I discovered in the course of my efforts to build the first production line, described in Chapter 6. However, they were wrong about the future of silicon technology.

They were not alone in their error. Germanium looked like the future to many transistor industry experts in the late 1950s.

Papers published around the tenth anniversary of the transistor, in a special issue of the *Proceedings of the IRE*, discussed the application of the device to computers and communications. Yet none of them so much as hinted at the revolution that silicon-based microelectronics would unleash in the 1960s. In fact, silicon transistors were not especially featured. William Shockley, who shared the Nobel Prize for the transistor, wrote:

It may be appropriate to speculate at this point about the future of transistor electronics. Those who have worked intensively in the field share the author's feeling of great optimism regarding the ultimate potentialities. It appears to most of the workers that an area has been opened up comparable to the entire area of vacuum and gas-discharge electronics ... It seems likely that many inventions unforeseen at present will be made based on the principles of carrier injection, the field effect, the Suhl effect, and the properties of rectifying junctions.[6]

The importance of the transistor would far outshine anything the experts of the early solid-state era could have imagined. None of them focused on the fundamental future importance of p-n junction devices in other applications.

## The big bang: The p-n junction

Smaller size, higher reliability, and much lower power requirements: these advantages would have made the transistor a logical successor to the vacuum tube even if it offered no other benefits. But the transistor used the concept of the p-n junction, which opened up endless possibilities.

The p-n junction is the basic building block of the transistor and most other semiconductor devices. It is the source of their functionality and flexibility. This remarkable device, which owes its theoretical foundations to the work of William Shockley,[7] ranks as one of the major scientific discoveries of the twentieth century.

It is well worth the effort to understand how it works, since everything that happens in digital electronics grows from its elementary functions.

### P-n materials in brief

The first step toward understanding solid-state electronics in general (and the p-n junction in particular) is to look at the properties of semiconductor materials, and explore how they can be altered to create certain electrical properties.

Semiconductors are either elemental – germanium and silicon, for example – or compounds, such as gallium arsenide and gallium nitride. In the pure state, as the name semiconductor implies, they exhibit electrical conductivity somewhere between that of insulators and metals, depending on the material.

---

[6] W. Shockley, "An invited essay on transistor business," *Proceedings of the IRE* 28 (1958), 954.

[7] W. Shockley, "The theory of p-n junctions in semiconductors and the p-n junction transistor," *Bell System. Technical Journal* 28 (1949), 435–489.

Add controlled amounts of certain other elements ("dopants"), however, and semiconductors gain very interesting and useful electrical properties. Depending on which element you add, a semiconductor will become either a "p-type" or "n-type" material.

Adding phosphorus atoms to silicon (or germanium), for example, produces an n-type material, possessing free electrons that can wander in the silicon's lattice structure, conducting electricity.

The addition of boron has the opposite effect and results in a p-type material. In this case there are too few electrons, creating "holes" (spaces for missing electrons) that similarly move freely through the crystal as conductors of electricity. We call freely moving electrons and holes *carriers*. Electrons, the carriers in n-type materials, carry a *negative* electric charge, while the holes found in p-type materials carry a *positive* charge.

## Building and using p-n junctions

The basic process of creating semiconductor devices consists of the controlled formation of n-type and p-type regions in close proximity. When we have adjoining p-type and n-type regions, we have a p-n junction.

What is special about p-n junctions? They are so versatile that just about every semiconductor electronic device which we will discuss, including transistors, light sensors and imagers, and light emitters such as semiconductor lasers, has one or more p-n junctions as core elements.

Figure 1.2 illustrates the electrical properties of the p-n junction. Wired one way (a), with positive voltage applied to the p-side and negative voltage to the n-side, the p-n junction is called "forward-biased," and acts as an electron injector from the n-side into the p-side, and a hole injector from the p-side into the n-side. In other words, it conducts electricity. The free electrons on the n-side are pushed by the negative voltage across the junction's barrier region into the p-side, while the positive holes are pulled into the n-side.

Forward-biased junctions are used as the *carrier injectors* in electronic devices, for example, as the "emitter" part of the transistor. In semiconductor laser diodes, a forward-biased junction provides the carriers that generate the light emitted by the laser.

Wire the p-n junction in reverse (b), however, with positive voltage applied to the n-side and negative voltage to the p-side, and it acts very differently. The positive voltage prevents the flow of current through the junction, making it "reverse-biased."

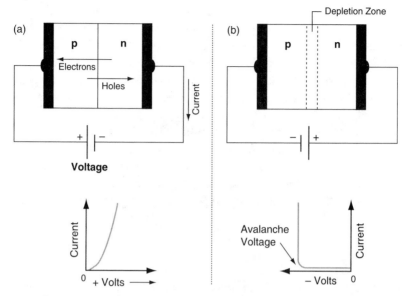

**1.2.** The two states of a p-n junction. (a) Forward-biased p-n junction showing current flow as a function of voltage; (b) reverse-biased p-n junction showing current being blocked until avalanche voltage is reached.

In this condition both electrons and holes are drawn away from the junction, creating a carrier-free region called the *depletion zone* on either side of it. By blocking the injection of carriers, the depletion zone prevents the flow of electricity through the device. The higher the reverse voltage, the bigger the depletion zone. Essentially the device is turned "off."

Increase the voltage beyond a certain value, however, and you get the whimsically-named *avalanche breakdown*, in which a sudden large current flows. At this point the high internal voltage near the junction in the depletion zone releases a stream of electrons from their tight atomic orbits, setting them free to move within the semiconductor. This effect does not damage the device.

This property is used in devices designed for controlling voltage levels in electrical circuits. Current flow is blocked until a critical voltage is reached, at which point the circuit begins to conduct electricity just as if someone had flipped an "on" switch.

Reverse-biased junctions are used primarily as *collectors of carriers* that can come from various sources. In the transistor, the reverse-biased p-n junction collects the carriers that originate in the emitter. In other applications (light sensors for example) a reverse-biased junction collects the carriers released from their atomic orbits by the incident light of the appropriate wavelength and this carrier flow produces a current flow in the junction. Devices based on p-n junctions are used to build imagers and light sensors of all kinds, including those used in fiber optic communications systems.

### From rectifiers to digital circuits

An important application of p-n junctions as they switch from the forward- to the reverse-biased state is as rectifiers of alternating voltage. When forward-biased, the device passes current; when reverse-biased it blocks it.

In effect, this is the first step in converting alternating to direct current in power sources. This is the basis of the power supplies that all electronic systems use to give their devices the direct voltage or current they need for their operation.

As will be obvious by now, the p-n junction is the enabling factor not just for rectifiers and transistorized amplifiers and switches, important as those are, but for many other essential devices.

Even more importantly, the p-n junction opened the way to creating extremely fast and efficient on/off current sources that can represent the ones and zeros of binary processing. It laid the foundation for creating miniaturized digital circuits.

## Applying the p-n effect: Bipolar transistors and MOSFETs

To see how p-n junctions enable transistors, let us look briefly at the two basic types of transistor, bipolar and MOSFET (metal oxide semiconductor field-effect transistor).

### Bipolar beginnings

The bipolar transistor was the first important commercial device of the 1950s. It consists of two p-n junctions back to back. The two n-type regions, called the *collector* and the *emitter*, are separated by a single thin p-type semiconductor layer, referred to as the *base*.

This structure, described in Appendix 1.1, is the successor to the original point-contact transistor invented at Bell Labs. The p-n junctions make it far more robust and practical than that first pioneering device. Instead of metal contacts forming the emitter and collectors, p-n junctions in the device perform this function. By removing metal contact points, we now have a completely solid-state device. So was born the modern *bipolar* transistor.

The bipolar transistor operates as a switch or an amplifier, but since the 1980s it has been gradually replaced by MOSFET structures.

### Advancing to MOSFETs

Although the first integrated circuits were built using bipolar transistors, since the 1980s they have been largely superseded by integrated circuits built around MOSFETs.[8] This device is the foundation of most of today's integrated circuits, memories, commercial transistors, and imaging devices.

Hundreds of millions of MOSFETs, miniaturized to the point that their size approach atomic dimensions, can be built and interconnected to fit on an integrated circuit chip the size of a fingernail. These microprocessors are at the heart of today's computers.

Like the bipolar transistor, the MOSFET also performs switching and amplifying functions. However, it is easier to manufacture on chips in high densities than the bipolar transistor. Even more important, a MOSFET dissipates far less power when it acts as a switch. However, its operation, also described in Appendix 1.1, is more complex.

Figure 1.3 shows, in schematic form, what a MOSFET structure looks like when built as part of an integrated circuit.[9] The important dimension to focus on is the distance L, which represents the length of the channel traversed by electrons when the transistor switches. The smaller L is, the faster the transistor switches, because the electrons have a shorter distance to travel.

This dimension is one of the key elements in determining the switching speed of the transistor, a parameter of vital importance in building computers. The smaller the MOSFET, the shorter L will be. Miniaturization of the transistor is therefore crucial in boosting speed. We will have a lot to say about miniaturization below.

---

[8] S. R. Hofstein and F. P. Heiman, "The silicon insulated-gate field effect transistor," *Proceedings of the IEEE* 51 (1963), 1190–1202. Also see D. Kahng, "A historical perspective on the development of MOS transistors and related devices," *IEEE Transactions on Electron Devices* ED-23 (1976), 655–657.
[9] J. R. Brews, "The submicron MOSFET," in *High-speed semiconductor devices*, S. M. Sze (ed.) (New York: A. Wiley-Interscience Publications, 1990), p. 144.

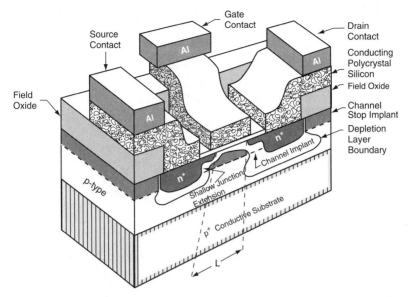

Gate Contact

Source Contact

Drain Contact

Conducting Polycrystal Silicon

Field Oxide

Field Oxide

Channel Stop Implant

Depletion Layer Boundary

Al

Al

Al

n⁺

n⁺

p-type

Shallow Junction

Channel Implant

Extension

p⁺ Conductive Substrate

L

**1.3.** MOSFET structure showing the various layers. The dimension L is the gate length, a key element in controlling device performance. Copyright © 1990 John Wiley & Sons, Inc. Reprinted with permission of John Wiley & Sons, Inc. (ref. 9).

Switching transistors on and off does not come without limitations. This function dissipates power, meaning the transistor heats up. The faster we switch, the more heat is generated and the hotter the chip gets. Heat must be removed from the chip in order to keep its temperature under control.

By way of compensation, however, the smaller we make the transistor, the less power it dissipates *individually* at a given switching frequency. Since the object of microelectronics is to make the transistor as small as possible, switch it as rapidly as possible, and dissipate minimal *total* chip power, this is a real benefit. It opens the way to putting more and more fast transistors on every chip.

### MOSFETs keep shrinking: Moore's "law"

In sum, there are many advantages to be gained by making MOSFETs smaller. We can put more transistors, and thus more processing capability, on a single chip. Their switching speed improves at smaller sizes. Appendix 1.1 discusses the relationship between the size of the MOSFET and its operating characteristics. As we have already noted, their individual switching power dissipation drops.

Gordon Moore, Intel's co-founder, observed in 1965 that the ability to shrink transistor dimensions was such that the processing capability of a single chip doubled approximately every two years. In effect, this doubled the performance of the chip even if chip costs remained constant (which they have not – costs keep dropping).

Table 1.2 shows how Intel was able to translate smaller transistors into ever higher performance for its microprocessors. As transistor dimensions continued to shrink, Intel put more devices on each new microprocessor, starting with 6,000 in 1974 and growing to 230 million in 2005.[10] The result was more processing power per chip. Progress is still being made: billion-transistor chips exist.

While transistor sizes were decreasing, as you would expect, speed was rising. Table 1.2 also shows the switching speed of the chips increasing from 4.77 MHz to 3.2 GHz during the same thirty-one-year period. Each generation of microprocessors has offered more processing power at higher speeds, achieved at ever-decreasing unit transistor costs.

Nature has been very cooperative in giving us the ability to shrink MOSFETs – but only up to a point.

**Limits to progress: Leakage current and thin connections on chip**
That point appears when we reach near-atomic spacing. As Figure 1.4 shows, the steady reduction in gate length L is putting us close to this mark.[11] It was 0.5 micron (500 nm) in 1993, and is projected to be only 0.032 micron (32 nm) in 2009. We are now facing the limits of our ability to shrink MOSFETs as currently understood.

There are two reasons why this has happened: excessive currents in the device as the source/drain p-n junction spacing approaches near-atomic dimensions; and the shrinking of the on-chip equivalent of interconnect wires. The metal interconnects between transistors cannot

---

[10] Data from H. Jones, IBS (2005), private communications.
[11] T. Thornhill III, CFA, "Global semiconductor primer," UBS Investment Research (UBS Securities LLC, an affiliate of UBS AG (UBS) – March 30, 2005), p. 82.

*Table 1.2 Intel microprocessor timeline*

| Microprocessor | Year introduced | Clock speed | Number of transistors (thousands) | Family groups |
|---|---|---|---|---|
| 8088 | 1974 | 4.77 MHz[a] | 6 | |
| 8086 | 1978 | 8 MHz | 29 | |
| 80286 | 1982 | 20 MHz | 134 | |
| 80386 | 1985 | 40 MHz | 275 | |
| 80486 | 1989 | 50 MHz | 1,200 | 66 MHz 1992, 100 MHz 1994, 133 MHz 1995 |
| Pentium | 1993 | 66 MHz | 3,100 | 100 MHz 1994, 133 MHz 1995, 166 MHz 1996, 200 MHz 1997 |
| Pentium Pro | 1995 | 200 MHz | 5,500 | |
| Pentium II | 1997 | 300 MHz | 7,500 | 400 MHz 1998 |
| Pentium III | 1999 | 500 MHz | 28,000 | 1.1 GHz[b] 2000 |
| Pentium 4 | 2000 | 1.5 GHz | 42,000–169,000 | 2 GHz 2001, 3 GHz 2002, 3.2 GHz 2003, 3.8 GHz 2004 |
| Itanium | 2001 | 800 MHz | 25,000 | |
| Itanium 2 | 2002 | 1 GHz | 221,000 | 1.5 GHz 2003, 1.6 GHz 2004 |
| Smithfield | 2005 | 3.2 GHz | 230,000 | Dual core |

*Notes:*
[a] MHz (megahertz)
[b] GHz (gigahertz)
*Source:* H. Jones, IBS (2005), private communications (ref. 10).

be reduced indefinitely in size without the risk of actually reducing switching speed.[12]

This has serious implications for future generations of microprocessors. To be specific, continuing to shrink gate lengths below about

---

[12] J. D. Meindl, "Beyond Moore's Law: The interconnect era," *Computing in Science and Engineering* 5 (January 2003), 20–24.

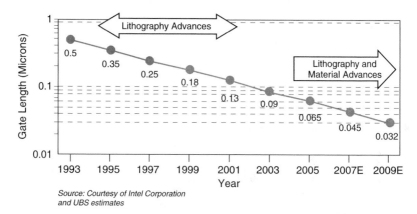

**1.4.** Gate length reduction over time. Lithography is the technology used to define the transistor feature size in manufacturing. Copyright © 2005 UBS. All rights reserved. Reprinted with permission from UBS (ref. 11).

0.065 micron (65 nm) is going to bring the interconnect problem to the forefront.

The most practical approach to solving the dilemma to this point has been the implementation of single-chip *multiprocessor* architectures. In effect, we build two or more interconnected computing devices on a single chip instead of one. This technique, discussed in more detail in Chapter 2, promises to let designers continue delivering advances in processing technology and architecture.

## Logic gates: Using transistors in digital processing

Up to now we have been considering the physical properties of semi-conductor structures and their impact on microprocessors. We've also seen how their structures allow them to perform basic electrical functions such as voltage control, amplification, and switching.

We're now ready to outline the basic principles of digital processing, and to show how they are embodied by the MOSFETs used to build computing chips. This is the turning point in modern electronics, when the electrical characteristics of miniature devices formed an exact match with the binary mathematics of digital calculation.

Digital computing uses Boolean algebra, which was developed in the nineteenth century as a tool for logical analysis. The Boolean system is the ultimate in logical simplicity. There are just two states: true or false. It is a binary system.

In the electronic universe, these two states are represented by the "bits" zero and one. They are produced by two different voltage levels, for example, zero volts and ten volts.

An electronic system can perform any binary arithmetic operation (or logic function) by using combinations of only three basic functional blocks, which are called *logic gates*. Logic gates control the path of the signals that execute the desired logical operations.[13]

The power of the binary system is that we can build up the most complex computers by simply replicating and interconnecting logic gates of only three basic types: NOT, AND, OR. To give a sense of the massive numbers of devices this can involve, the most complex CMOS microprocessor chips contain more than 100 million interconnected gates, each of which contains an average of four MOSFETs.

The ability to assemble the most complex computers by repeating such simple structures millions of times is truly remarkable and worthy of further discussion. Let's take a closer look at three basic logic gates, illustrated in Figure 1.5.[14] For clarity, we will treat the gates as switches that turn a light bulb on and off.

- An **Inverter**, or **NOT** gate (a), changes an input one into an output zero, and an input zero into an output one. If input signal to the switch is 1 (positive), as it is in the diagram on the left, the light is off (zero) because the signal is holding the switch open. When the input is zero, as is shown on the right, the switch is held closed by an internal spring and the light bulb is on (1).
- The **AND** gate (b) takes two or more inputs and produces a single output. This simplistic model shows two switches, A and B, in series. The output is zero and the light is off, as is shown on the left, if either or both switches are open; it is 1 and the light is on only if both of the inputs are 1, as it is on the right.
- The **OR** gate (c) produces a 1 when any of the inputs are 1, and a zero when all of the inputs are zero. Only when both switches A and B are off, as on the left, is the bulb off (zero). When either switch is open, as

[13] Z. Kohavi, *Switching and finite automata theory* (New York: McGraw-Hill, 1978).
[14] D. Cannon and G. Luecke, *Understanding microprocessors* (a Texas Instruments Learning Center Publication, 1979), pp. 3–13.

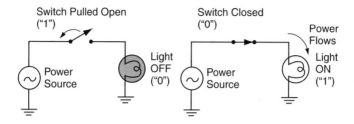

(a) Simple NOT Gate (Inverter)

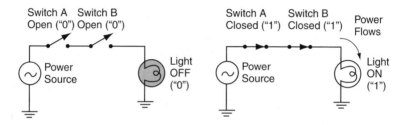

(b) Simple AND Gate

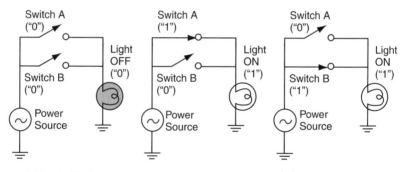

(c) Simple OR Gate

**1.5.** The three basic gates. (a) The NOT gate (inverter); (b) the AND gate; and (c) the OR gate. Courtesy of Texas Instruments (ref. 14).

in the middle and on the right, current flows to the light bulb, turning it on.

Digital logic also uses combinations of gates for its processing, such as the NAND gate. This is an inverter connected to an AND gate. Appendix 1.2 explains how such gates are produced with multiple transistors.

## *Power dissipation: The crucial role of CMOS*

We've noted that power dissipation issues are impeding progress toward further size reductions in transistors, a topic we will explore in more depth in Chapter 2. But this is actually the second time the semiconductor industry has found a power dissipation problem standing in the way of increased transistor density on chips – and solved it brilliantly.

The first time was in the early days of integrated circuit development. MOSFETs were already the obvious choice as building blocks for logic gates, but their architecture suffered from a serious flaw. Many of the transistors were always conducting electricity, even when they were not switching.

As shown in the example of a two-transistor inverter gate in Appendix 1.2, one of the MOSFET's devices was conducting current even when the other one was completely off and no output was required. The result was excessive power dissipation.

So while miniaturization of MOSFET transistors made it possible to put many logic gates on a microprocessor chip, too many of the transistors would dissipate power even at idle, creating serious overheating. This severely limited the number of transistors that could be used on a single chip.

The problem was solved with the invention of the complementary metal oxide semiconductor (CMOS) architecture, which reduced power dissipation enormously.[15] With overall power dissipation much lower, the door was open to massive increases in chip transistor density, from tens of thousands to many millions.

RCA introduced the first commercial CMOS integrated circuits in 1969. Practically all integrated circuits using MOSFETs are built with the CMOS architecture.

## Giving computers a memory

There would be no computers without the ability to store and rapidly retrieve massive amounts of digital information. The stored information can take the form of software programs or data.

---

[15] For a view of the early work on CMOS, see J. R. Burns, "Switching response of complementary-symmetry MOS transistor logic circuits," *RCA Review* 25 (1964), 627–661. Wanlass, US Patent 3,356,858, "Low power stand-by power complementary field effect circuitry" (December, 1967).

Software programs are stored instructions that tell a computer how to perform a task: word processors, financial spreadsheets, video editors, or wireless communications protocols. Data is the information that is created and manipulated during processing, such as documents, database tables, image files, and more.

Software and data must be stored within easy reach of the processor. There are three major media for storing information, each with its own advantages and drawbacks.

*Semiconductor memory* offers the fastest access because it is totally electronic. This benefit has a downside: all solid-state memories (except FLASH ones) will lose all the information they contain if cut off from electric power.

*Magnetic disks* store information on a physical medium where it is unaffected by power losses. They offer reasonably fast access, huge capacities, and very low cost per unit of storage. So-called hard disks are the primary magnetic storage devices, used in everything from desktop PCs to mainframes. Floppy disks have been supplanted by optical and solid-state storage. Magnetic tape is the medium for another class of magnetic storage systems. Tape is primarily used for storing backup information, an application where its much slower access time is less of a disadvantage.

*Optical disks* store data in a form that can be read by laser-based mechanisms. Originally read-only, they now include write-once and rewritable CDs and DVDs. They compete with magnetic media in that they provide permanent storage, but capacities are lower and optical access times are much slower. Their chief advantage is that the physical disks are removable, allowing for exchange of physical media and for inexpensive off-site storage of valuable data.

## MOSFET memory structures

Transistors are remarkably versatile. Nothing better illustrates this than the fact that, besides being used in logic gates for *processing* information, MOSFETs are the building blocks for data storage devices. These semiconductor devices are more commonly known as memory, computer memory, or simply RAM (for random access memory).

**Three memory types**

The most important class of semiconductor memory is dynamic random access memory (DRAM). In a DRAM a charge representing a bit of data is stored in a single transistor incorporating a special built-in capacitor. The charge on the capacitor needs to be constantly refreshed by special circuits. DRAMs offer the largest storage capacity for a given chip size precisely because they require only a single transistor per cell.

Static random access memories (SRAMs) constitute the second most important class of charge-storage memory. Instead of a single transistor with an attached capacitor to hold a charge, SRAMs use six transistors to create the unit cell for storage of a single bit. The basic concept is that this cell changes its state by having different transistors turned on or off depending on whether or not a bit is stored there. The master circuit on the SRAM chip can interrogate each cell and determine or change its status as required.

SRAMs do not need to have their stored charges refreshed. And while DRAMs have a storage density four times that of SRAMs for a given chip area, SRAMs offer important benefits, such as faster access time, that justify their higher cost in some systems.

DRAMs and SRAMs share one drawback: stored information is retained only as long as the devices are connected to a power source. This is not true of the third most important class of semiconductor memories, which is emerging into greater and greater importance.

Non-volatile memory (the term *FLASH memory* is more often used) will store cell charges in *practical* systems *indefinitely*, even without being connected to a power source. (Indefinitely does not mean forever – storage is reliable within a time period on the order of a few years.) This makes FLASH memory the first choice for such applications as retaining information in a cell phone, for example, which is frequently turned off.

Appendix 1.3 describes the basic principles behind the special MOSFET structure that makes up the unit cell of such memories.

Since MOSFETs have undergone a radical reduction in size over the past two decades, chip makers have been able to produce RAM chips with steadily increasing storage capacities at progressively lower cost. For example, four memory modules totaling 128 MB of RAM might have cost over $1,000 in 1990. In 2006 a single module that holds 1 GB, eight times as much memory, sold for under $100.

All memory devices in a digital system work on the same basic principle. They contain storage sites called cells, each of which can hold a "bit" of information. All that is needed from a memory cell is the ability to store (or not store) an electronic charge, which is equivalent to the presence or absence of a 1 or 0. It's binary logic at its simplest.

Each cell in a device, or in an array of such devices, has a unique address, called a memory location. A processor or other circuit can use that address to access the cell. It then performs either of two operations:

- interrogate the cell to obtain the value it is storing, for use in performing an operation;
- change the value as the result of a process or operation.

### Magnetic disk storage

The principle of magnetic disk storage dates to the mid-1950s, when IBM originated the concept of using tiny magnetic regions on the surface of a rotating disk as bit storage locations. The first useful product appeared in 1960.[16]

The surface of the disk is coated with a metal oxide, and the read/write/erase function is performed by an electro-magnetic head. A bit (one or zero) is written (stored) by a change in the magnetization of a tiny spot on the disk. Alternatively, the head leaves the bit in place but simply detects its presence.

The disk rotates at a speed in excess of 4,000 revolutions/minute. High-performance drives spin at more than twice that speed. The head never actually touches the magnetic material; instead, it floats very close to the disk surface.

The storage capacity of disk drives has doubled about every eighteen months since the late 1950s. This has been accomplished by finding ways to store a bit on a disk in less and less space, which translates into lower and lower cost per unit of storage.

Figure 1.6 shows the amazing decline in the cost of a megabyte of storage between 1958 and 2005.[17] The cost per megabyte has dropped by a factor of ten million!

---

[16] A. S. Hoagland, "History of magnetic disk storage based on perpendicular magnetic recording," *IEEE Transactions on Magnetics* 39 (2003), 1871–1875.

[17] Committee on the Fundamentals of Computer Science: Challenges and Opportunities, Computer Science and Telecommunications Board, and National Research Council of the National Academies, *Computer science: Reflections on the field, reflections from the field* (Washington, DC: National Academies Press, 2004), p. 90.

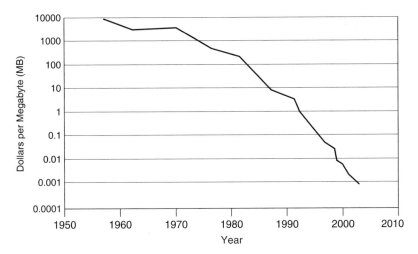

**1.6.** Cost per megabyte of magnetic disk storage over time. Reprinted with permission from *Computer science: Reflections on the field, reflections from the field*. Copyright © 2004 by the National Academy of Sciences, courtesy of the National Academies Press, Washington, DC (ref. 17).

This did not happen without overcoming serious technical challenges. As the heads put bits in smaller magnetic spaces on the disk, the magnitude of the signal in those spaces, which indicate the presence or absence of the bits, has gotten smaller as well.

Engineers have managed to design specialized integrated circuits to read these weak signals with improved reliability at ever-lower costs. Also, as the heads seek locations on the surface and perform their read/write/erase functions, sophisticated transistor-based electronic systems control their movement and manage the data flow.

Other key parameters of disks that have evolved over the years include faster access time, higher storage densities, and higher data transfer rates.

At one time, magnetic disks were literally the only practical permanent storage medium. Magnetic floppy disks dominated the removable storage market the way that hard disks controlled the fixed storage space. This is changing.

FLASH memories are competing with hard drives and floppy drives in lower-density portable device storage applications. The FLASH structure has an advantage, in that it is an all-solid-state device with no moving parts. In addition, the bit density on the chips continues to increase.

*Table 1.3 Price per gigabyte (GB) of storage (dollars)*

| Year | Disk drive | Flash memory | Price ratio |
|------|-----------|--------------|-------------|
| 2003 | 1.50 | 25.20 | 16.8 |
| 2004 | 0.81 | 12.15 | 15.0 |
| 2005 | 0.42 | 5.60 | 13.3 |

*Source:* Data collected by H. Jones, IBS (2005), private communications.

This is another effect of the shrinking of transistor dimensions, discussed earlier in this chapter. Chips with four gigabits of storage capacity became commercially available in the year 2006. Such chips contain about four billion transistors.

However, as of this writing (in 2006), FLASH memory is still over ten times as costly per unit of storage as magnetic disks. Table 1.3 tracks the price ratio per gigabyte between FLASH memories and hard disks. Although hard disks are still far less expensive, the difference is expected to narrow. It is felt that eventually the density of data on a disk will reach its limit, while FLASH densities will continue to rise.

Whether or not this happens, it is certainly true that FLASH memory, in the form of the plug-in USB "drive," has already replaced the floppy disk as the medium of choice in the personal removable storage market. Optical disks are making headway too, and are the only removable-media option for storing data in multi-GB quantities. However, their larger form factor puts them at a disadvantage in portable devices.

## Semiconductor laser diodes

Heterojunction semiconductor laser diodes are the devices that enable optical communications and many other systems. They power laser printers, read from and write to CDs and DVDs, enable certain medical procedures, and even provide the light source for the humble laser pointer so useful in public presentations.

We've grown so used to them that we no longer think about what the word "laser" really means. We just know what lasers do. (For the record, "laser" is an acronym for "light amplification by stimulated emission of radiation.")

Appendix 1.5 presents an introduction to the basic concepts behind laser diodes. For our purposes here we will focus on the fundamentals behind the enormous progress in semiconductor lasers since the late 1960s.

Semiconductor laser diodes, like MOSFETs, are everywhere, though there are far fewer of them. Like MOSFETs, semiconductor laser diodes have undergone a rapid reduction in price as their applications have proliferated.

Several factors combine to make heterojunction semiconductor lasers so useful compared to other light sources.

- Tiny size, reliability, and very low operating currents.
- Emission of *monochromatic* light energy. The emitted wavelength is determined by the choice of semiconductor and ranges from the blue into the far infrared.
- Ability to generate a single highly *directional* beam.
- Extraordinary internal efficiency (nearly 100%) in converting electricity into light.
- Ability to tailor properties and control costs accordingly from less than a dollar a unit to thousands of dollars for the most demanding requirements.

The heterojunction laser, which replaced earlier devices that performed poorly, was developed independently at RCA Labs,[18] the A. F. Ioffe Physicotechnical Institute in St. Petersburg, Russia,[19] and Bell Labs.[20,21]

---

[18] H. Kressel and H. Nelson, "Close-confinement gallium arsenide p-n junction lasers with reduced optical loss at room temperature," *RCA Review* 30 (1969), 106–114. The first commercial heterojunction lasers were commercially introduced by RCA in 1969.

[19] Zh. I. Alferov, V. M. Andreev, E. L. Portrisi, and M. K. Trukan, "AlAs-GaAs heterojunction injection lasers with a low room temperature threshold," *Soviet Physical Semiconductor* 3 (1970), 1107–1109.

[20] I. Hayashi, M. B. Panish, P. W. Foy, and S. Sumski, "Junction lasers which operate continuously at room temperature," *Applied Physical Letters* 17 (1970), 109–111.

[21] For a history of laser diode development, see H. Kressel and J. K. Butler, *Semiconductor lasers and heterojunction LEDs* (New York: Academic Press, 1977).

A heterojunction is the interface between two dissimilar semiconductors, whether they are p-type or n-type. In other words, it is a special case of the p-n junction that has figured so prominently in electronics since the 1950s.[22]

By using heterojunctions, it is possible to reduce the active lasing volume and hence the current needed to operate these devices. Using entirely different concepts from those applied to shrinking MOSFETs, this new architecture has made possible benefits similar to those of miniaturization.

Operating current levels are important in creating practical lasers. Figure 1.7 shows the historical reduction in the current density needed to sustain lasing in these devices.[23] Operating current density requirements have now dropped so low that a penlight battery is quite adequate to power a small laser diode emitting a few milliwatts of continuous red light.

Because of their low power dissipation and superior emission characteristics, it is possible to modulate specially constructed heterojunction-based laser diodes at gigabit/second rates at room temperature. This makes it possible to create optical communications links operating at those rates. However, the highest data rate systems (up to 40 Gb/s) use an electro-optic modulator between the laser diode and the fiber to produce the optical bit stream.

Nor is their power dissipation a significant problem in fiber optic systems, although temperature control is provided to keep the lasers operating within their optimum temperature range. Other major improvements have been also made in the operating characteristics of laser diodes to adapt them to many applications.[24]

Figure 1.8 shows in schematic form what these devices look like.[25] Note the light beam emitted from a small region which, in this simple

[22] H. Kromer, "A proposed class of hetero-junction injection lasers," *Proceedings of the IEEE* 51 (1963), 1782–1783.

[23] Adapted from Kressel and Butler, *Semiconductor lasers*, p. 4.

[24] Two publications offer interesting accounts of these advances: M. Dutta and M. S. Stroscio (eds.), *Advances in semiconductor lasers and applications to optoelectronics* (Singapore: World Scientific, 2000) and "Special issue on semiconductor lasers," *IEEE Journal of Selected Topics in Quantum Electronics* 9 (September/October, 2003).

[25] H. Kressel, I. Ladany, M. Ettenberg, and H. Lockwood, "Light sources," *Physics Today* 29 (May 1976), 45.

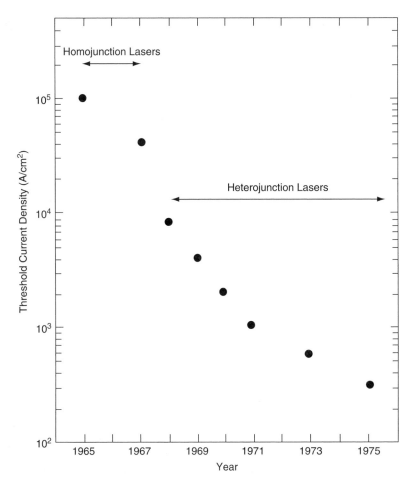

**1.7.** Historical trend in the gallium arsenide and heterojunction aluminum gallium arsenide laser diode threshold current density at room temperature. Technological improvements after 1975 focused on heterojunction lasers of indium gallium arsenide phosphide emitting at 1.55 microns for fiber optic systems. Reprinted from *Semiconductor lasers and heterojunction LEDs*. Copyright © 1977, with permission from Elsevier (ref. 23).

structure, is only 10 microns wide. Figure 1.9 shows how tiny laser diodes are when fully packaged as a general purpose laser emitting a beam through the front glass platelet. A device this size can emit hundreds of watts of pulsed power!

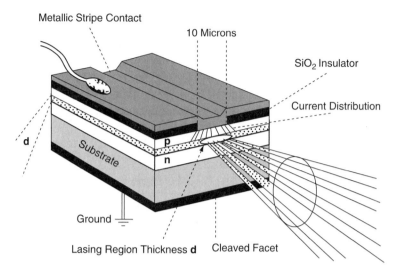

**1.8.** Schematic of a simple double heterojunction laser diode structure showing light beam emitted under stripe contact 10 microns wide. The front and back of the structure constitute the two surfaces of the Fabry-Perot cavity. Reprinted with permission from *Physics Today*. Copyright © 1976, American Institute of Physics (ref. 25).

### A key requirement for systems: Device reliability

We should not leave the subject of the rapid progress made in developing practical electronic devices without reflecting on what it tells us about the gap between new device concepts and their practical utility. Perhaps the most important lesson we can learn has to do with device reliability.

Practical consumer electronic systems can tolerate device operating lifetimes of tens of thousands of hours. This is not the case for communications systems. In these applications predicted device lifetimes must be well in excess of ten years. For undersea communications links, lifetimes have to be a lot longer.

The early history of all electronic devices, and particularly of laser diodes, was plagued by erratic lifetimes. A thousand hours of life was noted as a great achievement. This problem was rarely discussed in the published papers that described the wonderful properties of a new structure.

In fact, it was not unusual for researchers to report test data on a new device without mentioning that the device lasted just long enough to

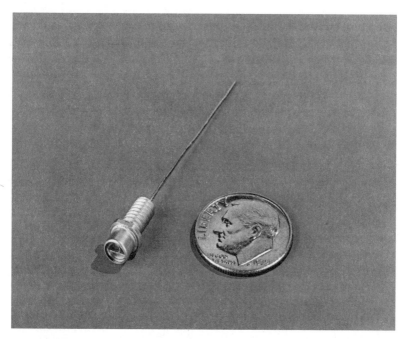

**1.9.** A laser diode packaged in a metal case with a glass face for the emitted radiation.

make the measurements. For example, early MOSFETs degraded rapidly because of impurities, such as sodium, in the oxide layers. Laser diode degradation was a complex phenomenon which turned out to be largely due to crystalline defects that increased in density as the device operated, eventually leading to its destruction.[26]

Driven by the importance of commercial value, the work at RCA Labs and other commercial laboratories was heavily focused on improving device reliability. It took years of patient work by a number of laboratories to create devices that offered the many years of reliable operation needed in practical systems. This work involved painstaking research on atomic-level effects of temperature, stress, and operating currents.

We take it for granted that "solid-state" means "forever" when compared to vacuum tubes. This only proved to be the case after many years of first-class research and development.

[26] Kressel and Butler, *Semiconductor lasers.*

## Optical fibers and modern digital communications

An optical communications system uses lasers, photodetectors, and glass fibers to transport digital data in the form of short light pulses over very long distances. Of the three components in this basic system, the only one not built around solid-state electronics and p-n junctions or heterojunctions is the transport medium, the optical fiber.

We have already explored the transformational technology of laser diodes, and Appendix 1.7 offers a brief primer on photodetectors. But no consideration of the innovations that formed the digital computing and communications landscape would be complete without including the development of optical fibers.[27]

Simply put, the infrastructure that enables wideband digital communications exists only because of glass fibers. Their extraordinary ability to carry data at high speeds over long distances is the reason we have the Internet. It is also why such high-capacity data transmission services as video are possible over the Internet.

### *The glass backbone of digital communications*

Up to now we've been focused on digital data primarily in the form of electrical signals, routed through electronic devices such as transistors and integrated circuits. Optical fibers, however, are pipes that carry data not as electricity, but as bursts of laser light.

The data originates as rapidly alternating voltages representing streams of digital bits. The voltages turn a laser diode on and off very quickly, producing laser pulses that faithfully replicate the ones and zeros of the data. The optical pulses enter the fiber optic cables. At the receiving end, p-n junction light sensors regenerate the original electrical pulses, which provide input to signal processing systems.

The concept and early demonstration of using fiber optics for communications over long distances was explored in the mid-1960s in the UK at the Standard Telecommunications Laboratory of ITT,

---

[27] For a good succinct historical overview, see T. E. Bell, "Fiber optics," *IEEE Spectrum* (November 1988), 97–102. For a longer history, see J. Hecht, *City of light: The story of fiber optics* (New York: Oxford University Press, April 1999).

as reported in a publication by Dr. Charles Kao and Dr. George Hockam.[28]

Commercial optical fibers trace their origins to the late 1960s, when Corning scientists invented thin glass fibers of extraordinary purity that could faithfully and reliably transmit short optical pulses over long distances.[29] This achievement required many innovations on the part of the scientists, plus some fortunate but unplanned coincidences.

In one such coincidence, nature helped solve a crucial difficulty. The scientists discovered glass compositions that happened to absorb very little light at precisely the near-infrared wavelength (1.55 microns) where InGaAsP heterojunction laser diodes were designed to operate.[30]

Figure 1.10 shows how optical fiber transmission loss was reduced over a ten-year period.[31] Between 1973 and 1983 signal loss due to light absorption was reduced from 100 dB (decibels) per kilometer to under 0.2 dB per kilometer (at about 1.55 microns), approaching the theoretical limits set by the properties of glass.

This low value, combined with the development of excellent InGaAsP heterojunction laser diodes emitting in the appropriate spectral region, enabled fiber links many kilometers long that could carry high data rates with acceptable transmission loss.

Another problem involved sending short pulses over long distances with minimal distortion. In a plain glass fiber, an input pulse (representing a bit of information) will spread out as the light propagates through the fiber. Eventually the shape of the pulse flattens out. This makes it impossible to determine that a bit of information has been transmitted.

The answer was to design the refractive index profile of the glass fiber to produce an internal "lensing" effect that keeps the pulses coherent. Of course, this property had to be consistent with low-cost

---

[28] C. K. Kao and G. Hockham, "Dielectric fiber surface waveguides for optical frequencies," *Proceedings of the IEEE* 113 (1966), 1151–1158.

[29] For a review of this development see R. D. Maurer, "Glass fibers for optical communications," *Proceedings of the IEEE* 61 (1973), 452–462. A survey of recent progress can be found in a special issue of the *Journal of Lightwave Technology* 23 (2005), 3423–3956.

[30] B. E. A. Saleh and M. C. Teich, *Fundamentals of photonics* (New York: John Wiley & Sons, Inc., 1991), pp. 592–643.

[31] C. Kumar and N. Patel, "Lasers in communications and information processing," in *Lasers: Invention to applications*, J. H. Ausubel and D. Langford (eds.) (Washington, DC: National Academies Press, 1987), p. 48.

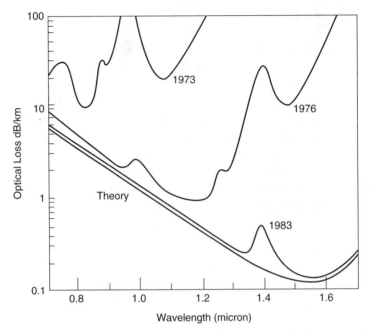

**1.10.** Spectral loss data for silica optical fibers over time. The heterojunction lasers used in communications systems emit in the 1.55 microns region. Reprinted with permission from *Lasers: Invention to applications.* Copyright © 1987 by the National Academy of Sciences, courtesy of the National Academies Press, Washington, DC (ref. 31).

mass production techniques. Appendix 1.8 discusses this aspect of the technology.

It was also necessary to develop cladding techniques that protect the delicate fibers as they are deployed over land or undersea. All these objectives were met, and the fiber optic industry was born.

The cost of fiber optic cables has dropped radically over the last two decades. This is the result of high-volume fiber production in huge plants. Figure 1.11 tracks the price per meter of single-mode fiber since 1983.[32]

During this period the cost of the semiconductor components used in optical communications was drastically declining as well. The

---

[32] Figure courtesy KMI Research, www.kmiresearch.com.

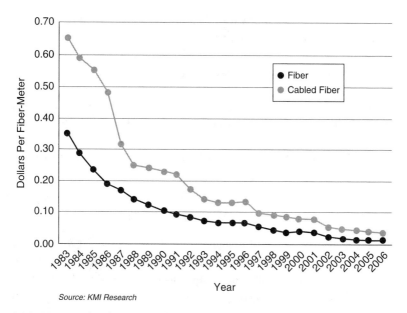

*Source: KMI Research*

**1.11.** Historical reduction in the cost of a fiber-meter of single-mode optical fiber. Figure courtesy of KMI Research, www.kmiresearch.com (ref. 32).

combination of plummeting system costs combined with increasing demand drove the global deployment of fiber optic communications.

Now, just a quarter-century after low-loss fiber was first demonstrated, universal availability of high-speed communications for voice, data, and video is practically taken for granted. The world will never be the same.[33]

## Semiconductor imagers and displays: Visualizing information

We began this chapter by recounting the transition from vacuum tubes to solid-state devices. Our focus since then has been on digital innovations for processing, storing, and communicating information: transistors, integrated circuits, lasers, and more.

But humans are analog. We can't handle digital data in electronic form the way a computer chip can. We actually need to see the input and output of the process.

---

[33] "Technologies for next-generation optical systems," special issue of the *Proceedings of the IEEE* 94 (2006), 869–1035.

Here, too, electronic innovation has created more efficient solutions. The cameras we use to capture the physical world are now completely solid-state. Vacuum tube-based video camera imagers have been replaced by semiconductor imagers.

On the display side, the bulbous CRT picture tube, derived from a device invented by Karl Ferdinand Braun in 1897 and used in televisions for over half a century, is quickly giving ground to flat-panel alternatives.

To close out our survey of the devices that underlie the digital world, then, we will briefly survey the technology behind solid-state imagers and displays.

## *Digital image capture: CMOS camera sensors*

The original semiconductor imagers used charge-coupled device (CCD) technology, which requires unique manufacturing processes. CCDs are still used in many high-end digital cameras and camcorders, but they are now facing a serious challenge from imagers built with CMOS-based technology.

CMOS imagers were once confined to applications where quality was less important than price. For the past few years, however, they have offered image quality every bit as good as all but the best CCDs. They still maintain their cost advantage, too, because they can be manufactured on standard IC production lines (with some modifications).

This repeats the pattern followed by all other digital semiconductor devices: steadily increasing performance at a rapidly decreasing price. The technology that enables this progress is the by-now familiar MOSFET structure.

Basically, an imager consists of an array of photodiodes that generates an electric signal proportional to the intensity of the light that strikes it.[34] Each photodiode represents one pixel (picture element), the dots that make up the final image. Figure 1.12a compares the CMOS approach to that of the CCD.[35]

In Figure 1.12a, the CCD architecture appears on the left. The photodiodes sense light, generate electrical signals in response, and

---

[34] Information on CMOS imagers is adapted from A. El Gamal and H. Eltoukhy, "CMOS image sensors," *IEEE Circuits and Devices* 21 (2005), 6–12.

[35] D. Litwiller, "CMOS vs. CCD: Maturing technologies, maturing markets," *Photonics Spectra* (August 2005), 55.

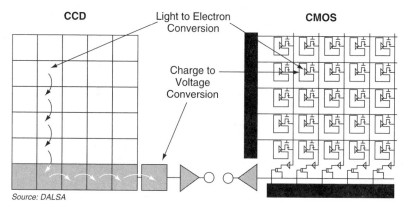

**1.12a.** Two types of imagers. Left, CCD and right, CMOS-based with MOSFETs incorporated into pixels. Reprinted with permission from D. Litwiller, DALSA (ref. 35).

transmit the signals (electronic charge) via a "bucket brigade" pixel-to-pixel transfer process. This is an analog method of transmitting the signal level (i.e., the number of electrons generated in a pixel) to the periphery of the chip, where the signals are processed.

On the right we see a CMOS-based imager. Each pixel has its own processing circuitry, consisting of MOSFETs, the output of which is sent directly to the periphery of the chip, and from there to the camera system.

Figure 1.12b is a diagram of the elements that constitute a digital camera.[36] As indicated, the CMOS imager chip actually incorporates many of the functions of a camera system. With a CCD, all of these functions have to be implemented separately. CMOS obviously represents a potentially large cost advantage to the camera manufacturer.

The block labeled "color filters" merits some attention. Regardless of the type of imager, the photodiodes (or pixel sites) that convert the detected light into an electric charge are monochrome. In order to image color pictures, the chip has to use three adjacent light sensors, one each for red, green, or blue light, for every imaging unit. For this reason the chip is covered with a regular pattern of red, green, and blue filters, so that each element of the triad can respond to just one color.

---

[36] Litwiller, "CMOS vs. CCD," 56.

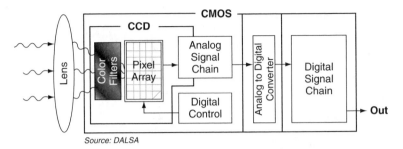

Source: DALSA

**1.12b.** Architecture of simple imaging system using a CCD imager. Reprinted with permission from D. Litwiller, DALSA (ref. 36).

The circuitry at the chip periphery receives the current generated by incident light on each color of pixel, and does the processing that allows the color image to be reconstructed on a display.[37]

To add to the complexity of this process, the resolution of the image is a function of the number of pixels on the chip. This can be as high as eleven million pixels on a single chip about one centimeter square.

Fortunately, this is achievable with today's technology. The CMOS production process only needs a 0.130 micron (130 nm) minimum feature size to achieve this density. Many microprocessors and memory chips are manufactured at much smaller dimensions. In the case of imagers, however, it's actually better to have larger features, because photodiodes collect less light as their size decreases.

Thanks to all of these advances, CMOS imagers are not only good enough to be used in still and video cameras of all kinds, they are inexpensive enough to be incorporated in wireless handsets.

It's interesting to note that more cameras are now sold as part of cell phone handsets than as stand-alone products. Over five hundred million wireless handsets will eventually be equipped with solid-state camera chips.

*Display technology: Amorphous silicon and organic transistors*

If there's an imager in use – or a processor – there's a display waiting to show the results. Increasingly that display is solid-state.

---

[37] Litwiller, "CMOS vs. CCD," 54–60.

A technical discussion of liquid crystal displays (LCDs) appears in Appendix 1.9. For our purposes in this chapter, however, the most interesting aspect of displays is the materials from which they are made.

So far our discussion on semiconductors has dealt with crystalline forms. These followed an historical evolution from germanium, to silicon, to compound semiconductors whose special properties gave them particular value in electro-optic devices.

But two other classes of materials have found their way into large-scale device applications: amorphous silicon and polymers.[38] These materials, produced as thin films, find their most important applications in solid-state flat-panel displays. We will consider amorphous silicon first.

**Amorphous silicon and LCDs**
In contrast to crystalline materials, which have well-ordered atomic arrangements, amorphous materials can be thought of as similar to glass, where atomic arrangements are only of short order. The resulting lack of a regular form (such as the lattice structure of a crystal) is the source of the designation *amorphous*.

The crystalline silicon wafers on which integrated circuits are built do not lend themselves to integration with materials such as the liquid crystals used for displays, or for use as a means of producing inexpensive electronic devices. It would obviously be desirable just to deposit a thin film of silicon on glass (at a temperature that did not melt the glass) and then build display circuits from that.

When research was begun to develop such a process in the early 1980s, it was found that the silicon deposited at low temperatures is actually an *amorphous* film. It had interesting applications, although it could not serve as replacement for the "classic" silicon crystals used to build integrated circuits.

The first application of amorphous silicon was in solar cells (for solar power generation). These devices could be built cheaply, helping to make up for their much lower efficiency compared to standard silicon crystals.

It was also discovered that one could fabricate large arrays of "inferior" transistors with this material. These transistors exhibited very

[38] P. Yam, "Plastics get wired," in a special issue of *Scientific American: The solid-state century* (1997), 91–96.

slow switching speeds, yet were still useful in limited applications far removed from microprocessors. Techniques to improve the electrical quality of the thin silicon films include heating them to improve their structures.

Thin-film silicon electronics were the key to the commercial manufacturing of large flat-panel liquid crystal displays.[39] The thin-film silicon devices incorporated in the display provided just the right level of pixel switching capability to make practical displays at acceptable cost.

The LCD flat-screen display opened up brand new applications, such as laptop computers, that were totally impractical with CRTs (picture tubes). Once again solid-state devices provided a more practical alternative to vacuum tube technology.

Figure 1.13 shows that the turning point in the market transition from tube-based to flat displays occurred in 2002.[40] The figures cited include all display-based devices, including PDAs, cell phones, computers, and TVs.

Even in television receivers, formerly a bastion of CRTs because of their lower price point, the dominance of the tubes is rapidly fading as LCDs and other semiconductor displays take hold. Figure 1.14 shows the dramatic price reductions for large LCD TV receiver displays that began in 2003, turning them from a luxury item into a common consumer product.[41]

### Polymers and OLEDs

Polymers are emerging as important devices for special applications in displays, light emitters, and even solar power generation (solar cells).[42]

---

[39] For a review of amorphous silicon properties and applications, see M. S. Shur, H. C. Slade, T. Ytterdal, L. Wang, Z. Xu, K. Aflatooni, Y. Byun, Y. Chen, M. Froggatt, A. Krishnan, P. Mei, H. Meiling, B.-H. Min, A. Nathan, S. Sherman, M. Stewart, and S. Theiss, "Modeling and scaling of Si:H and poly-Si thin film transistors," *Materials Research Society Symposium Proceedings* 467 (1997), 831–842.

[40] *Special technology area review on displays*, a report of Department of Defense, Advisory Group on Electron Devices, Working Group C (Electro-Optics) (Office of the Under Secretary of Defense, Acquisition, Technology and Logistics, Washington, DC, March 2004), p. 16.

[41] H. Jones, IBS (2005), private communications.

[42] For display applications see H. Hoppe and N. S. Sariciftci, "Organic solar cells: An overview," *Journal of Material Research* 19 (2004), 19–24. Applications to solar cells are reviewed by S. R. Forrest, "The path to ubiquitous and low-cost organic electronic appliances on plastic," *Nature* 428 (2004), 911–920.

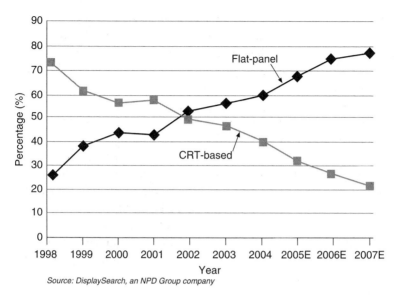

**1.13.** Flat-panel display sales surpassed cathode-ray tube (CRT) sales in 2002. Reprinted with permission from DisplaySearch, an NPD Group company (ref. 40).

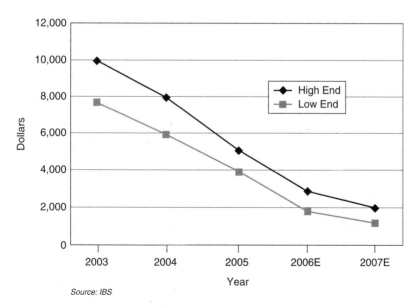

**1.14.** Price of 45-inch LCD TV display over time. Reprinted with permission from IBS (ref. 41).

In the course of investigating other kinds of semiconducting materials, researchers discovered that certain polymers could be used to construct thin-film transistors with limited properties.[43]

It was also discovered that polymers could be used to create thin-film light-emitting devices with emission in the blue, green, and red, suitable for manufacturing flat-panel displays. These are generally called OLEDs (organic light emitting diodes), though they are known as organic EL (electroluminescent) displays in Japan.[44]

The ability to use polymer-based electronics is expected to have a far-reaching impact on applications such as large-area displays. With flexible substrates replacing glass as the base material, large displays will become more portable, less fragile, and less expensive.

Other advantages of OLEDs include lower power consumption and potentially lower production costs. Several companies are racing to develop a method for printing the transistors and diodes on a continuous roll of thin plastic, which could enable "video wallpaper" at extremely low prices.

## Foundations for digital systems

The proliferation of new electronic devices that started in the late 1940s was the first wave of technology in the flood tide of digital products that still engulfs us. These fundamental components washed away previous assumptions about how people could create and use technology to interact with information.

The major impact of these devices wasn't their ability to miniaturize products that had previously been unwieldy, power-hungry, and limited in functionality. It wasn't the potential to achieve higher performance at drastically lower prices either, though that was certainly important.

It was the promise of huge, powerful systems built on tiny elements, rather like living organisms composed of invisible atoms. Each of these devices plays a role in creating the immense power of the computing

---

[43] H. Sirringhaus, N. Tessler, and R. H. Friend, "Integrated optoelectronic devices based on conjugated polymers." *Science* 280 (1998), 1741–1744.

[44] J. H. Burroughes, D. D. C. Bradley, A. R. Brown, R. N. Marks, K. Mackay, R. H. Friend, P. L. Burns, and A. G. Holmes, "Light-emitting diodes based on conjugated polymers," *Nature* 347 (1990), 539–541.

and communications systems that affect human life in every corner of the world.

In Chapter 2, we will see how innovation at the system level changed the world to a digital environment and helped to create the global village. It all started, however, with a handful of tiny devices, developed in corporate labs long ago.

# 2 | *Building digital systems*

B Y the early 1970s, most of the basic electronic development that made digital technology feasible was in place. One great burst of innovation over two decades had produced the transistor, the p-n junction, the CMOS process for integrated circuits, magnetic disk storage, heterojunction semiconductor lasers, fiber optic communications, and semiconductor imagers and displays.

The speed with which researchers put the new devices to practical use was almost as remarkable. They quickly developed an array of systems that gather, process, and transmit information.

The most visible example was the computer. In 1982, *Time* magazine focused popular attention on the computer's importance by naming it "Machine of the Year," a widely discussed departure from the "Man of the Year" honor the magazine had bestowed for over sixty years.

Since then computers have become a part of our lives. Smaller, more powerful microprocessors and supporting devices, combined with plummeting costs, have sparked an endless stream of new computerized consumer and industrial products.

However, while the computer has been central to the Information Age, the most revolutionary transformation has taken place in communications. Data transmission, once limited to 300 bits per second on analog phone lines, now uses digital links to carry multi-megabit per second data streams. Voice has gone digital and wireless. The Internet has metamorphosed in just ten years from an "insider" network for scientists to an indispensable, ubiquitous communications backbone linking billions of people around the globe.

And both computing and communications have been enabled by advances in software architecture.

This chapter summarizes the development of computing systems; examines wired and wireless communication for voice, data, and video; and explores the software innovations that endow hardware with intelligence and purpose.

56

It also looks at how the explosive growth of digital systems is changing the business and economic landscape, especially in communications and software. Customers are no longer locked into telephone companies for voice service, and companies now have alternatives to proprietary software.

Cheap computing and universal high-speed communications have an impact on society which is far beyond the spheres of technology and industry. This chapter lays the groundwork for our later exploration of the socioeconomic implications of the digital revolution.

## Computing systems: Data processing made affordable

Personal computers are now so commonplace that it's easy to forget how exotic – and forbiddingly expensive – they once were. In 1990, a desktop PC with an 8 MHz processor, 128 Kbytes of memory, and an arcane command-line operating system cost $4,000. Fifteen years later a pocket-sized personal digital assistant with a user-friendly touch-screen interface offers more memory and a speedier, more powerful processor for $99.

This astonishing progress is typical of the digital revolution: as products become smaller, faster, more powerful, and easier to use, they also get cheaper – much cheaper. It's no accident that during the past decade computing has become a normal activity, even in underdeveloped parts of the world.

None of this would have happened if electronics had not made the leap from analog to digital. Analog electronics were not a practical platform for large-scale computing and communications systems. Readers interested in a technical explanation for this will find it in Appendix 2.1, "The demise of analog computers," which also describes how digital systems interact with the analog world.

For our purposes, the important thing to keep in mind is the scalability of digital technology. Scalability allows designers to put high functionality into tiny devices, then assemble them into powerful, relatively inexpensive computing systems.

Since the microprocessor is at the center of all modern digital systems, we will start by focusing on its technology and applications. Later in the chapter, when we look at the communications and software industries that the microprocessor made possible, the true extent of its economic impact will become apparent.

## Microprocessors: Affordable processing power

It's worth repeating that digital computation is conceptually simple. In Chapter 1 we saw that combinations of only three basic logic gates (NOT, AND, and OR) are needed to perform any logic or arithmetic function. All you need is a processing unit with enough gates and adequate access to memory. The more gates you have, the faster and more powerful the processor becomes. This scalability is the biggest advantage of digital technology.

Until the 1970s, getting enough gates was a costly proposition. The central processing units (CPUs) in computers were built around silicon integrated circuits containing custom-designed logic gate arrays. This made them very expensive. The situation changed dramatically with Intel's introduction of the microprocessor in 1971.

Gordon Moore, co-founder of Intel, saw the microprocessor as an alternative to building logic for each individual application. He described the device as "a broadly applicable, complex integrated logic circuit that can be produced in huge volume, and hence [at] low cost, and that utilizes the technology advantageously."[1]

Experts of that time expected that microprocessors would be used to control traffic lights, manage automobile functions, and perform other mundane applications.[2] These predictions have proven accurate.

However, nobody envisioned the major role of microprocessors: enabling ever more powerful, ever cheaper PCs. And no one foresaw that the mainframes of the 1970s would be largely replaced by microprocessor-powered computers.

Moore's forecast of huge production volumes has been fulfilled beyond anything he could have predicted. Today microprocessors are everywhere, at all levels of performance. There are hundreds at work for each human being in the world.

They coexist with many other types of digital and analog/digital chips to make up complete systems. Figure 2.1 shows sales of semiconductors worldwide since 1999, broken out by type of chip and total revenues.[3]

---

[1] G. E. Moore, "Microprocessors and integrated electronics technology," *Proceedings of the IEEE* 64 (1976), 837–841.
[2] A. J. Nichols, "An overview of microprocessor applications," *Proceedings of the IEEE* 64 (1976), 951–953.
[3] Data from H. Jones, IBS (2005), private communications.

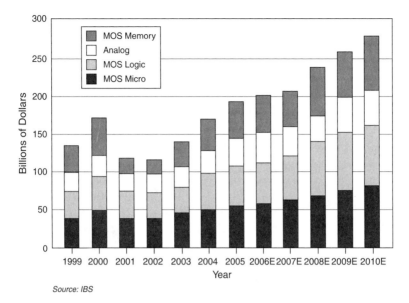

**2.1.** Integrated circuit market segments between 1999 and 2005, and projected to 2010. Microprocessors are included in the "MOS Micro" segment (from IBS, ref. 3).

Microprocessors command an important share, yet other types of ICs exhibit strong sales as well.

The great majority of these ICs use MOSFETs, described in Chapter 1, to enable digital and analog functions. Because of this they benefit from the same progress in manufacturing technology that has steadily increased the power and reduced the size and cost of microprocessors.

### Flexible, scalable processing

While the other IC types are vitally important, it's still the microprocessor that established the dominance of digital electronics. In one form or another it is at the heart of every digital product sold today.

The scalability of its architecture in terms of instruction size makes it the ideal device for digital computation. Designers can choose a microprocessor with the level of power and sophistication they need, at a price point that makes sense for the final product.

Depending on its power, a microprocessor can vary in cost from under a dollar for an 8-bit unit to hundreds of dollars for the high-end

64-bit unit used in the most complex systems. Each of these devices contains the major elements of a computer, small or large.

As the speed and performance of microprocessors have improved, and their cost has dropped, they have found their way into everything from toys to washing machines to large computers. For example, over fifty microprocessors are found in every automobile,[4] and that number is rising (see Chapter 8).

Not all of these devices are full-function, stand-alone products such as those sold by Intel and AMD. Many are "cores," smaller processors with limited functionality embedded on a larger chip.

A core provides the computing engine for ICs containing a system on a chip (SoC). Designers use SoCs to carry out tightly defined applications, such as managing the data flow in cell phones.

A stand-alone 64-bit microprocessor with fifty million gates is clearly overkill for such a simple application, and it draws too much power. A 32-bit core embedded in an onboard SoC chip, on the other hand, may have only 100,000 gates. That's enough power to do the job, with little additional battery drain, and it costs much less than the stand-alone IC.

### When smaller is not better

Every new generation of microprocessors has offered higher performance at a lower price, often in a smaller package. That progression has driven the decline in system costs from $4,000 PCs to $99 PDAs (personal digital assistants). But how long can these dramatic advances continue?

Moore's Law, described in Chapter 1, predicted that the processing power of ICs would double every two years with no increase in price. This was based on the rate at which MOSFETs were being shrunk in size (Appendix 1.1), permitting more logic gates on each chip. As Table 1.2 showed us, market leader Intel has achieved dramatic increases in chip densities over the years.

Today, however, MOSFETs are approaching the practical limits of miniaturization. Making them smaller once meant getting more speed

---

[4] R. Kline, "An overview of twenty-five years of electrical and electronics engineering in the *Proceedings of the IEEE, 1963–1987*," *Proceedings of the IEEE* 93 (2005), 2170–2187.

for less money. But as MOSFET gate dimensions shrink well below the micron (100 nm) level, and the density of transistors on the chip continues to increase, we begin to bump up against physical constraints on performance.

In addition, as we observed in Chapter 1, the connections among IC components have become so thin that they are slowing on-chip communications. More ominously, the combination of increased density of transistors on the chip and higher clock rates (speeds) is generating severe power dissipation problems. This has the potential to raise the operating temperatures of chips above a safe level.[5]

Figure 2.2 shows how both static and switching sources of power dissipation have evolved in advanced processors.[6] Since 2002, when 0.09 micron (90 nm) gate spacing became practical, static current leakage from the rising number of transistors on a chip became more significant as a source of power dissipation than switching frequency. (This assumes that the transistors are switched at their theoretical limits.)

With clock rates exceeding 3 GHz,[7] a single high-end microprocessor can easily surpass 100 watts of power dissipation. For most applications, a value of thirty watts or less is needed, and portable devices require considerably lower values to prevent excess battery drain. At 100 watts it's also hard to keep the device from overheating without water-cooling.

Microprocessor architects are learning to control power dissipation by designing the chip so that only part of the device is functioning at any one time. Other sectors stay in stand-by mode, waiting for activation as needed. Designers also minimize power dissipation by mixing different size transistors on a chip, using the smallest and fastest devices only in critical areas. New materials can also help. But these approaches are only partial solutions.

---

[5] A comprehensive review of this situation can be found in J.P. Uyemura, *Introduction to VLSI circuits and systems* (New York: John Wiley & Sons, 2002), pp. 257–259; and R. Goering, "Leakage takes priority at 65 nm," *EE Times* (January 16, 2006), 47–48.

[6] J. Gea-Banacloche and L. B. Kish, "Future directions in electronic computing and information processing," *Proceedings of the IEEE* 93 (2005), 1858–1863.

[7] From Gea-Banacloche and Kish, "Future directions in electronic computing," 1858–1863.

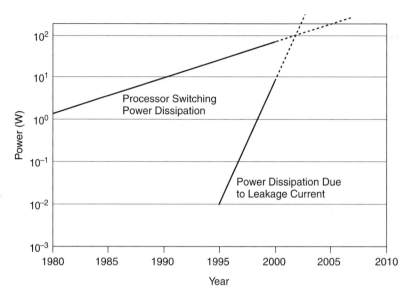

**2.2.** Extrapolation (dashed lines) of processor switching power dissipation and leakage current-related power dissipation. Fit line replotted (solid line) and extrapolated (dashed line) with G. Moore's presentation in 2003 (see ref. 3). The crossing of the extrapolated lines indicates that the leakage current power dissipation is progressively dominant from around 2002–2003. Copyright © 2005 IEEE (from Gea-Banacloche and Kish, "Future directions in electronic computing," ref. 6).

### Breathing room: Parallel processors

It appears, then, that the strategy of squeezing more performance and lower cost out of a microprocessor by continuously shrinking the size of transistors in order to increase their speed and density is reaching its physical limits. However, there is another answer: parallel processors.

Parallel processing has long been done by building a single system around multiple discrete processors.[8] Now, however, Intel, AMD, and others are using well-known parallel processing concepts in single-chip, multi-core implementations.

In these configurations each core has its own processing resources, but shares access to centralized control functions, including instructions

[8] See J. L. Hennessy and D. A. Patterson, *Computer architecture – a quantitative approach*, 3rd edition (Amsterdam: Morgan Kaufman Publishers, 2003), pp. 636–659.

and data storage. Since the individual processors have significant autonomy in the execution of instructions, this arrangement reduces the requirement for high-speed on-chip communications. The problem of processing speed being slowed down by the thin metal lines between transistors is alleviated as well.

For operations where parallel processing is easily implemented, a dual-processor chip can deliver about 70 percent more throughput than a single-processor chip running at the same clock rate (speed). The price disparity between the two chips is very small. By anyone's standards, that's an impressive boost in price/performance ratio.

### The problems with parallel processing

Parallel processing is neither simple nor a panacea. To achieve its full benefit, the software has to be designed to take advantage of the subdivision of computing tasks into parallel tracks. This works well for such applications as the analysis of data packets in networks and images. Other computing tasks are more difficult to divide, and the benefits of using parallel processing for these applications are less clear.

In addition, it can be difficult to convert software programs designed for single microprocessors to a multi-core environment.[9] It is estimated that 100 billion lines of code exist for the X86 Intel microprocessor family. Conversion of all this software may never happen, but proponents of parallel processors argue that new software designs are inevitable anyway if computing costs are to continue to decline at historical rates.[10]

There is little doubt, on the hardware side at least, that parallel processing is the wave of the future. Published reports suggest that Intel expects that 40 percent of all desktop computers, 70 percent of all mobile devices, and 85 percent of all servers produced after 2006 will use multi-core processor chips.[11]

So at least for the foreseeable future each new generation of computing systems will continue to be smaller, more powerful, and cheaper, thanks to a combination of reduced feature size and parallel processing.

---

[9] K. Krewell, "Software grapples with multicore," *Microprocessor Design* 19 (December 2005), 1–4.

[10] P. Hester, CTO of AMD, quoted by D. Lammers, "New microarchitectures, from the ground up," *EE Times* (January 16, 2006), 20.

[11] R. Kay and P. Thibodeau, "Counting cores: the newest processors cram two or more CPUs onto a single chip," *Computerworld* (June 20, 2005), 48–54.

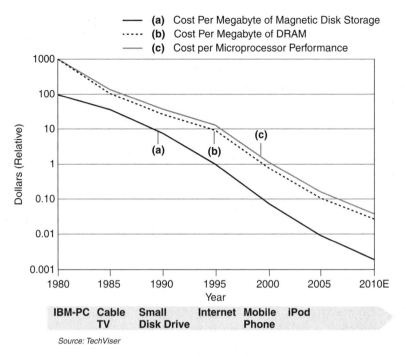

**2.3.** Historical relative cost of three elements of computing systems: magnetic disk storage, DRAM, and microprocessor performance. Reprinted with permission from Mashey, TechViser (ref. 12).

## Computing hardware is (nearly) free

Computers are everywhere, in and out of sight. Costs have dropped so low that embedding computer controls in a new product is often less costly than using older hydraulic or mechanical technology. As a bonus, a computer-based product can provide additional features and functions.

As Figure 2.3 demonstrates, the price of computing power has declined by more than four orders of magnitude since 1980 and should continue to drop.[12]

The graph breaks out cost reductions for the three major components of a computing system: disk storage, DRAM memory, and microprocessors. All three track quite well over time. As we've seen before, much

---

[12] J. Mashey, TechViser (2005), private communications.

of the historical drop in microprocessor costs to date can be traced to progressive reductions in transistor feature sizes. The same is true for DRAMs and the chips attached to magnetic disk storage devices.

Going forward, multiprocessor chips promise to help microprocessor performance to increase at declining prices. Cost reductions in other devices, such as DRAMs, will also come from new architectures and further shrinkage of MOSFET dimensions.

In the case of magnetic disk storage, future cost reductions will come from perpendicular recording, a new magnetic technology that yields a 45 percent increase in storage density on the disk surface compared to the current method. Perpendicular recording differs from the traditional longitudinal method in that it aligns data bits vertically rather than horizontally on a disk platter.

Seagate Technologies, the leading provider of hard disks, started marketing 2.5 inch perpendicular storage disks with 160 GB capacities for portable use in early 2006. A few months later it announced its first desktop models, boasting 750 GB capacities for $560 at introduction.

As other disk makers follow, further progress in increasing densities is sure to come. Eventually it may be practical to use optical holograms to increase storage densities.

Figure 2.3 also suggests something interesting about the way computer technology responds to the pressures of supply and demand. The dates when important computer-related products were released, listed along the bottom axis, point to a correlation between lower prices and increased demand. Evidently, when demand goes up, supply goes up even more.

## Communications systems: Wideband information access

Computers made digital communications possible. But digital communications has turned out to be the more disruptive technology. The social and economic effects of its widespread deployment are just beginning to be felt. A look at its advantages over earlier models will prepare us to understand why it is having such a major impact.

### *Advantage digital: Speed, distance, flexibility*

Like computers, modern communications must be digital. Had communications remained in the analog domain, the Internet would not exist, and neither would computer networks, cell phones, or even the modern telephone system.

The reason, as we indicated in Chapter 1, is that high-speed transmission of large amounts of analog information over long distances is simply impractical. Different parts of an analog signal will travel at different speeds over copper lines or cables. As the distance increases, the signal shape grows increasingly distorted, and signal strength drops. Eventually information gets lost.

In principle, the problem can be solved if we first digitize the information and transmit it as pulses. In optical communications systems, we use light pulses to transmit information. A pulse would correspond to a one, and the absence of a pulse in a time slot to a zero. We can then detect the presence or absence of pulses at the receiving end of the transmission link.

While the pulses can distort or be reduced in intensity as they travel, no information is lost as long as we can differentiate one pulse from another at the end of the path. It is also easy to maintain the signal strength of digital communications by reformatting and amplifying the bit stream at intervals along the way. This makes long, high-speed fiber optic links practical. For example, fiber links 10,500 km long can theoretically transmit reliably at 40 Gb/s.[13]

Of course this is an oversimplification based on fiber optic networks, but it helps in understanding the principles of digital communications. In actual practice, the use of pulse trains applies only to fiber optics. When digital data is transmitted over coaxial cables, through copper lines, or wirelessly, the bits are encoded on parallel electrical sine waves by highly complex methods based on sophisticated computations. Thanks to increasingly powerful microprocessor chips, however, this technology is also extremely reliable.

Digital communication offers overwhelming advantages over analog:

- A single network can be used for all services. Because binary digital transport is content-neutral, it carries data, voice, or video equally well.

---

[13] M. Lefrancois, G. Charlet, and S. Big, "Impact of very large cumulated dispersion on performance of 40Gbit/s submarine systems over nonzero dispersion fibres," *Electronics Letters* 42 (2006), 174–176.

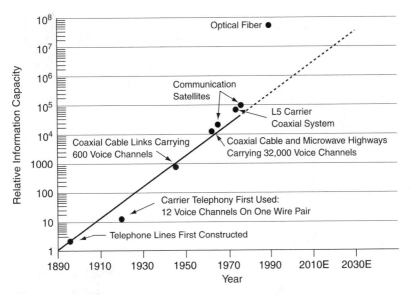

2.4. US national communications channel capacity improvement as a function of time. Note the discontinuous change in the average slope with the introduction of optical fiber systems. Reprinted with permission from *Lasers: Invention to applications*. Copyright © 2004 by the National Academy of Sciences, courtesy of the National Academies Press, Washington, DC (ref. 14).

- Speed and capacity are far higher. Advances in signal processing, spurred by declines in computing costs, continue to produce dramatic increases in bit rates.
- Compressed data streams, possible only with digital technology, yield substantial increases in capacity.
- Fiber optic transmission, practical only with digital data, has produced phenomenal increases in the information-carrying capacity of networks, as shown in Figure 2.4.[14]
- International standards defined by organizations such as the Institute of Electrical and Electronic Engineers (IEEE) and the International Telecommunications Union (ITU) have made worldwide digital system interconnection practical.

[14] C. Kumar and N. Patel, "Lasers in communications and information processing," in *Lasers: Invention to applications*, J. H. Ausubel and H. D. Langford (eds.), (Washington, DC: National Academies Press, 1987), p. 47.

**2.5.** AT&T telephone switchboard operators (property of AT&T Archives. Reprinted with permission of AT&T).

## *Switched connection, data overload*

To understand how disruptive the digital communications revolution actually is, let's compare the original public switched telephone network (PSTN), which prevailed for a century, with its successor, packet-switched digital systems.

The original analog telephone system completed a call by establishing temporary connections between subscribers. In the earliest days, when Mabel wanted to call Joe, she would alert the operator. If the line was free, the operator would connect Mabel's line with Joe's on an early manual switchboard, and get out of the way. When the call was completed and both parties hung up, the operator was signaled that both lines were free for another call.

This manual process, with an operator directly involved in every connection, is shown in Figure 2.5. As the number of lines and calls grew beyond the capacity of operators to handle them, the process was automated in three successive phases.

- Electro-mechanical switches set up the call, interconnected the subscribers, and then disconnected the callers. They also managed all dial and ring tones needed during the process.
- Automated electronic switches ultimately displaced the electro-mechanical ones in the PSTN. The first electronic switches were analog, but digital models began appearing in the 1980s. Traffic was still analog.
- Eventually digital transmission, using the concept of time-division multiplexing (TDM), replaced analog in the network backbone. TDM carried high-bit-rate data streams among its many communications channels through dedicated time slots.

TDM-based digital transmission seemed so advanced. No one expected that in a few years data would far surpass voice as a percentage of network traffic, but that's just what happened. Voice networks were never built to address that eventuality. Soon the PSTN was facing a crisis.

To grasp the dimensions of the crisis, consider what happens if we want to carry data over a TDM switched network. While a (voice) telephone call lasts only a few minutes, a data call can keep lines connected for hours. This places a huge burden on the switches that have to maintain those connections.

The phone companies responded by setting up dedicated data lines. These are always-on connections which bypass the voice switches and link selected destinations on a permanent basis. It was a reasonable solution for large companies with heavy data traffic needs. To this day businesses use leased lines to connect geographically dispersed facilities where alternatives do not exist.

For smaller companies and individuals, however, this was not a workable alternative. These users, whose data communications volume did not justify the expense of a leased line, continued to burden the PSTN with data traffic. They needed a more flexible network architecture to handle their data requirements.

## *The connectionless IP network*

In a stroke of irony, the solution was developed in answer to an entirely different set of needs: those of the US Department of Defense (DoD), a large user of data communications services. The DoD understood how vulnerable a switched network was to wartime devastation. Its strategists started looking for redundant networks that would bypass destroyed nodes and allow military traffic to survive in the event of an attack.

Researchers hit upon the idea of a connectionless packet-switched network. Connectionless packet networks are built on the radical concept that you don't need a dedicated traffic path to distribute digital data.

Instead of formatting a continuous stream of information that must be kept intact from origin to end-point, it's possible to divide the information into individually addressed digital "packets" of data. These packets can find the way to their destination independent of each other. Even if they're broken up *en route*, they can be reassembled into a complete transmission when they arrive.

Though complex in its execution, the process is easy to understand.

- The packets traverse a network consisting of data transport links such as fiber optic lines.
- Routers (specialized computers) are placed at nodes in the network to read the packets and direct them to the next node until they reach their destination.
- Optical signals are converted into electrical ones at each node, to allow the routers to do their work.
- The routers read the address headers on the packets, compare them to lookup tables stored at the node, determine the best available path to the destination, and send the packets on their way.
- The packets are held at their final destination until the whole transmission arrives, after which the complete data transmission is put back together for the recipient.

This basic concept is attributed to Paul Baran, who was at the Rand Corporation when he published the architecture in a study for the US Air Force in 1964. Later it became the basis of ARPANET, developed under sponsorship of the Defense Advanced Research Projects Agency (DARPA). ARPANET eventually evolved into the Internet.[15]

By 1977, ARPANET was a network of 111 computer switching nodes providing data services primarily to research centers and

---

[15] A useful history can be found in Kline, "An overview of twenty-five years," 2170–2187; and J. Naughton, *A brief history of the future: The origins of the Internet* (London: Weidenfeld & Nicolson, 1999), pp. 93–97. A technical description will be found in P. Baran, "On distributed communications networks," *IEEE Transactions on Communications Systems* CS-12 (1964) 4, 1–9.

government facilities. The first commercial packet-switched network, TELENET, formed by contractors who operated ARPANET, began service in 1975.

While PSTN networks for voice traffic will be around for a long time, packet-switched networks are now universally accepted as the path to the future. A huge amount of work has been done to standardize their operations.

### How packets work

A packet consists of two parts:

- the *payload*, a group of bits that contains the data content to be delivered;
- the *header*, containing the packet's origin and destination addresses, plus instructions for its handling.

The header does the real work. The payload is along for the ride.

Header information uses rigidly prescribed protocol standards, established by the networking industry, which allow packets to be universally recognized. To get to their destinations, for example, packet headers use the Internet Protocol (IP) addressing system, described in more detail in Appendix 2.2, "IP, TCP, and the Internet." IP is the *lingua franca* of the networking business.

But the header contains more than simple routing information – much more. In fact, header instructions that ensure packets are properly treated as they travel to their destination can account for as much as half of a packet's total bits.

This is necessary because of the random nature of a digital packet network. The packets in a single transmission from one point of origin can be routed through different nodes, and may arrive at their destination out of sequence.

Some data, such as live video or voice, do not lend themselves to being delivered out of sequence. That's where the headers come into play. The headers of packets that need to travel together can be tagged with a special code telling a router to keep them together *en route*.

Packet networks use standard protocols to tag packets that require such special treatment. The universally accepted tags make it possible to deliver time-sensitive data packets to their destination in the same sequence and with the same time allotments they had at their origin.

Appendix 2.2 also contains a more detailed explanation of how this is accomplished.

## Packet-switched networks enable new services

Packet-switched networks, especially the Internet, are changing the way we communicate, and shaking up the established order of the world's communications industry in the process. To summarize:

- The networks are flexible. Different services can be provided by managing packet headers and provisioning lookup tables on the routers. This allows services to be handled the right way during transport. When people speak of "intelligence at the edge of the network," this is what they mean: computers at the edge of the network tag packets and provision routers with updated tables to enable a variety of defined services to be implemented by packet management.
- If the network is provisioned by appropriate protocols to handle time-sensitive classes of packets in their proper sequences, it does not matter whether we transmit data, voice or video. Voice and video become just two more data types.
- The reliability of these new networks is very high. Instead of using a dedicated voice network based on a switched architecture, the traffic travels over a widely dispersed network consisting of interconnected routers. Because the routers are independent engines, the network has built-in redundancy. If one segment goes down, packets can be rerouted to their destination through alternate paths.

The speed, flexibility, and reliability of digital communications, from the simplest text e-mail to video on the Web, are a direct outcome of the packetized network structure.

## The World Wide Web (WWW)

The unique accomplishment of the Internet is its ability to seamlessly interconnect hundreds of millions of computers on the basis of a standard addressing scheme. Its extraordinary value arises from two attributes:

- nearly universal transport properties;
- protocols enabling efficient sharing and distribution of documents, plus the ability to extract data from within documents.

In other words, almost any type of information can be carried by the Internet; and it can all be shared by anyone with a computer and access to the network.

The second of these attributes became far more practical in 1989, when Tim Berners-Lee and Robert Cailliau at the Centre Européen pour la Recherche Nucléaire (CERN) in Geneva designed the hypertext system for encapsulating data. Hypertext refers to the way information can be shared. It allows each page stored on an accessible site to link to another page. Using hypertext, information retrieval or search is a process of finding connected links in response to a query.

Hypertext addressed the need for convenient data and document sharing among scientists using the Internet. As the first protocol for Web markup, Hypertext Markup Language (HTML) formed the basis of the World Wide Web's document sharing capabilities.

But being able to share documents isn't enough. You must first find them. Since HTML is relatively primitive, with limited search capabilities, the Web needed a mechanism to enable intelligent data search. This was done by adding more information to the hypertext "tags" in Web documents.

The protocols for handling documents have since evolved into the now-dominant eXtensible Markup Language (XML), which defines the digital notation for content "tagging" and presentation of data.[16] XML formats include standard definitions that allow a wide range of industries to define their products. For example, the computer industry has standard XML descriptions so that when someone who wants to buy a PC conducts an Internet search, it will quickly find comparative data on computers offered for sale.[17]

Further enhancements continue to extend XML's capabilities in managing document searches on the Web. The Semantic Web project, under way as of this writing, promises to generate improved standards for adding information to the document tags. The overall aim is to improve the ability to extract data from within documents.[18]

---

[16] XML was adapted from the Standard Generalized Markup Language (SGML), long used in industries such as aerospace and pharmaceuticals to catalog complex documents with extensive specifications. SGML defines formats for documents.

[17] R. Vidgen and S. Goodwin, "XML: What is it good for?" *Computing and Control Engineering* (June 2000), 119–121.

[18] R. Kay, "Semantic Web," *Computerworld* (February 27, 2006), 32.

## Managing the wealth

By one estimate, as of February 2005 there were 11.5 billion indexed pages on the World Wide Web.[19] Searching all these documents is obviously impossible without innovative concepts in data management and the document search process, and several prominent search engines have filled this gap. We will cover this subject in detail in Chapter 8.

A more fundamental issue is how to direct traffic among the Web sites that house all this information. Since 1998, the Internet Corporation for Assigned Names and Numbers (ICANN), under control of the US Department of Commerce, has been responsible for the Internet. There is a central hardware architecture, consisting of thirteen powerful backbone computers that route traffic worldwide. Four of these computers are in California; six are near Washington, DC; and there is one each in Stockholm, London, and Tokyo.

ICANN also supervises a system of domain names that provides the organizing principle for this traffic. An Internet subscriber's computer is assigned an IP address, which consists of a unique number. When it accesses the Internet, the traffic is routed through computers in the Domain Name System (DNS) that translate the number into a name address (such as XYZ.com). Subscribers, of course, see only the name address.

ICANN appoints the managers of the DNS. For example, addresses ending in .com are managed in the computers of the Verisign Corporation, while .biz addresses are managed by the Neustar Corporation. Everything else is handled behind the scenes by the computers in the backbone.

The Internet, including the World Wide Web, now has over one billion users worldwide. Some authorities project that Web traffic will grow at 100 percent per year through 2015.[20] Built in just a decade, the Internet ranks as one of the greatest and most profoundly transforming feats of technology in history. Chapter 8 will discuss its commercial and social impact.

---

[19] A. Gulli and A. Signorini, "The indexable Web is more than 11.5 billion pages," a research paper posted at www.cs.uiowa.edu/~asignori/web-size/ (accessed on March 27, 2006).

[20] Briefing to Warburg Pincus by Morgan Stanley Research, April 16, 2006. Original source: Cisco Systems.

## Enabling IP phone calls

We've seen how packet networks took the bulk of data traffic away from the PSTN. Now IP-based systems are threatening to do the same with voice traffic.

One of the new services enabled by IP-based packet networks is IP telephony, or voice over IP (VoIP). VoIP became a factor in the market in 2004, when it reached the million customer mark. Although digitized voice and video signals are transmitted just like any other data, and problems relating to time-sensitive information flow had largely been solved years before, other issues had to be addressed before VoIP networks could match the PSTN in voice quality.

For one thing, there was the difficulty of reproducing the sound of the human voice. At first glance this might not seem like much of a problem. While human hearing is sensitive to a much wider frequency spectrum, the bandwidth of human speech is mostly between 600 and 6000 Hz. The bit rates required to encode this sonic range are well within the capacity of the network.

But encoding is complicated by other properties of individual expression, including variations in pitch, frequency, and intensity. Furthermore, the ear is very sensitive to sound variations. It can pick up delays lasting just a few milliseconds. The digitizing process had to be designed very carefully to prevent voice users from having unpleasant hearing experiences.

To allow digital processing of voice information, sound waves of varying frequency and intensity are converted to digital signals (see Appendix 2.1 for a description of the digital sampling process), and then packeted. The voice packets use header protocols that request special treatment to prevent delays and other audible artifacts.

At this point the packets are ready to be carried as VoIP calls. This involves converting the digitized signal to a TCP/IP bit stream, then setting it up through a special computer in the network, which usually uses the Session Initiation Protocol (SIP) to handle signaling, caller identification, connection, routing, and disconnection.

This sounds complex, but it actually streamlines the handling of calls. Because they use the Internet, VoIP calls do not require a hierarchical set of expensive switches. This gives VoIP a cost advantage over conventional PSTN voice transmission.

## The VoIP challenge to the phone company

Low cost is not VoIP's only attraction. Its IP functionality gives VoIP a number of advantages in the commercial market, including:

- Unlimited nationwide calling for a flat per-month fee, at rates competitive with, and sometimes less than phone companies charge for local connections only.
- Free caller services, such as voicemail and caller ID, which are usually extra-cost options from the phone companies.
- Portable phone numbers: a subscriber can keep the same number, even if he or she moves from New York to San Francisco.
- Inexpensive international calls.
- Location-independent telephony: subscribers can make and receive "local" calls on their VoIP phone number from anywhere in the world by plugging their adapters or, on some services, their PCs, into a broadband connection.
- Web-enabled features, such as getting messages by e-mail or on the Web, which are simply unavailable with standard phone service.

With IP-based voice systems supplanting traditional phone systems in many corporate communications systems, VoIP is already transforming the enterprise telephony business, forcing changes in the telephone companies' business models. The carriers are beginning to offer VoIP services of their own in self-defense, but at the risk of reducing their revenues per minute of call.

The VoIP tide is rising in the consumer market as well. With call quality improving and services becoming more widely available, a growing number of tech-savvy individual customers are jettisoning their standard service in favor of VoIP links. As Figure 2.6 shows, US subscribers to VoIP telephony are expected to rise from 1.1 million in 2004 to 22 million in 2009.[21]

The shift to IP-enabled voice communications has profound implications for the telecommunications industry. Since the industry derives much of its income from telephony, VoIP threatens a major revenue source.

This is because, historically, a major part of the telephony industry's revenue stream is based on billing phone calls by the minute. Data

---

[21] P. Cusick, CFA, "NeuStar, Inc.: Thriving on telecom turmoil" (Bear Stearns & Co., Inc., Equity Research – August 26, 2005), p. 14.

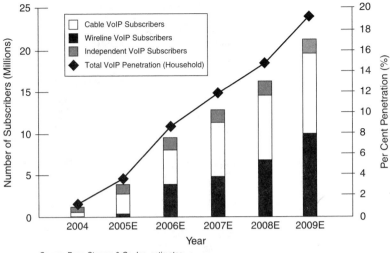

Source: Bear, Stearns & Co. Inc. estimates
For illustrative purposes only

**2.6.** Recent and anticipated growth of VoIP customers in the US. These numbers are most likely understated given the VoIP service of Skype and of companies such as Google and Yahoo! Reprinted with permission from Bear Stearns (ref. 21).

connections, on the other hand, are assumed to be always on, and are generally billed at a flat fee based on data rate.

For example, a one-minute digital (not IP-based) phone call requires 480,000 bits, for which a phone company will charge $0.02. This doesn't seem like much, but it is actually an artificially high price. The call represents a relatively small data stream which, in principle, should cost no more than the transport of any other data stream of that size.

By comparison, a 2 Mb/s Digital Subscriber Loop (DSL) connection to the Internet from the same phone company may cost a flat $20 a month. A short mathematical exercise shows that if a one-minute call were made over an IP service bundled with that DSL service, the cost per minute of access works out to only $0.0005 – almost nothing.

What's more, since this call uses only a fraction of the data capacity of the DSL line, subscribers can continue to use the line for e-mail, Web browsing, or other activities without affecting the call.

A VoIP call is not totally immune to network tolls. Calls can incur costs from conventional telephony access providers, who charge an access fee when a call jumps onto the PSTN to reach the called party's location. However, VoIP carriers usually offer "free" VoIP-to-VoIP calls among their subscribers. Since these never touch the conventional network, their cost is covered by the monthly subscription charge.

**Skype moves VoIP off the phone**

To see how digital technologies can transform whole industries, consider the impact of a venture capital-financed start-up company which emerged in 2003.

Skype offers voice service based on a proprietary protocol capable of excellent call quality. It differs from standard VoIP services like Vonage in several respects.

- Unlike existing VoIP services, it requires no central servers. Instead, it uses subscribers' computers/telephony gateways to provide connectivity.
- While services such as Vonage use stand-alone adapters to connect phones to IP-based networks, Skype is fundamentally a PC-to-PC service (although a new option called SkypeOut can provide PC-to-phone connections).
- Vonage and similar services assign standard phone numbers to subscribers. Skype subscribers do not receive a phone number unless they subscribe to the SkypeIn option.

In many ways Skype works like a peer-based music-sharing service. There's a good reason for this. Its founders, Niklas Zennström and Janus Friis, were the founders of KaZaA, a leading music site.

There were PC-to-PC phone schemes before Skype, but they were unreliable, and offered erratic (and frequently very poor) voice quality. Skype achieves good call quality through a relay technique, routing calls through other subscribers' equipment to bypass network bottlenecks. Strong encryption, integrated Instant Messaging, and secure file transfer complete the offering.

Just as they did for music sharing, Zennström and Friis have taken PC-to-PC voice communication out of the hobbyist realm and made it commercially acceptable.

Skype's revenue model is based on charges recovered when a call is made to a conventional PSTN number using SkypeOut. Skype-to-Skype calls are free, and there is no subscription or setup fee.

The technology attracted a lot of attention when it was announced, which helped Skype sign up an estimated 100 million subscribers around the globe in just over two years.[22] Between three and five million subscribers are on line at any one time.

The company also attracted a huge sales price from online auction giant eBay, which acquired Skype in September 2005 for €2.5 billion in cash and stock.

In a development entirely characteristic of new digital technologies, formidable competition quickly entered the market. Google, Earthlink, Lycos, and Yahoo! have all announced similar services.

**The economic effects of change**
VoIP technology can interface with any transmission medium. As shown in Figure 2.6, its adoption rates are increasing among all types of voice telephony customers, including those using wireline telephones, cable connections, or independent (Internet access) services.

Phone service has long been the beneficiary of government regulation and artificial price controls. But when voice calls are just another data service, these barriers crumble. The impact of VoIP on the revenues of established public carriers promises to be significant, and the ready availability of a cheap communications service will change social structures too, particularly in developing nations.

In Chapter 8 we will consider how all digital technologies, including VoIP communications, transform existing industries.

## *Digital video over IP networks*

Video is one of the fastest-growing applications in the information society. Here, too, packet-based networks are a natural medium.

For example, as broadband connections become more common, video is popping up all over the Web. In fact, according to CacheLogic, a provider of traffic monitoring and management technology to Internet service providers, in 2004 peer-to-peer file-sharing traffic already accounted for 60 percent of Internet traffic, with video making up 62 percent of that amount.[23] This is in addition to the growing use of streaming video on commercial Web sites.

---

[22] Briefing to Warburg Pincus by Morgan Stanley Research, April 16, 2006.
[23] "CacheLogic technology installed in tier-one ISPs worldwide monitors actual network traffic, reveals surprising facts in audio/video trends and format usage" (August 9, 2005), Press Release accessed on May 23, 2006 at www.cachelogic.com/news/pr090805.php.

But transmitting digital video over an IP network poses an even greater challenge than voice. It requires handling data from sequential frames made up of anywhere from 300,000 to two million pixels, for standard and high definition resolutions respectively. The signal contains an enormous amount of information – color, sound, and intensity.

In fact, the data stream of a standard US broadcast television program at 30 frames/second, using 8 bits/pixel to describe its parameters, is approximately 140 Mb/s.

To reduce the demands that video places on the network, providers normally compress this signal. Data compression relies on sophisticated mathematical algorithms for encoding the image at the source and decompressing it for the destination display.[24]

The algorithms compress the image by encoding information that describes the full content of the image every few frames, but describes only the changes in the image for the intervening frames. The decoder at the receiving end reconstructs the partial frames in such a way that the viewer sees little or no deterioration in the image.

Industry standards are essential if video systems from various manufacturers are to work together. The Joint Photographic Experts Group (JPEG) has developed standards for still images that can produce a compression ratio as high as 20:1.

For motion video the most widely accepted standards come from the Motion Picture Experts Group (MPEG). MPEG-2, for example, is the standard used to compress data for DVDs, the US broadcast digital TV system, and most digital satellite television. The newer MPEG-4 Part 10 standard, more commonly called H.264, offers higher compression ratios and more features. Satellite services are adopting it for high-definition video content. It is the choice for next-generation video applications, including video on cell phones.

These standards, combined with reduced frame rates and lower resolution, have compressed video to data streams as low as 64 Kb/s. While this is adequate for some applications, it is not a TV-quality viewing experience.

---

[24] For a review of theory and methods, see R. Steinmetz and K. Nahrstedt, *Multimedia: Computing, communications and applications* (Englewood Cliffs, NJ: Prentice Hall, 1995).

Under MPEG-2 a data stream of about 1 Mb/s (compressed) is required to transmit VCR-quality video. An MPEG-2 encoder will compress the 140 Mb/s stream of a standard-definition signal to between 1 Mb/s and 5 Mb/s. It will encode a 1 Gb/s HDTV signal at 10 Mb/s to 18 Mb/s.

As compression algorithms continue to improve, we may see video at much higher resolutions become readily available in multiple streams on the Internet.

## The wireless world

Guglielmo Marconi first demonstrated long-distance radio transmission in 1897. A century later consumer and industrial wireless applications had exploded, thanks to dramatic improvements in semiconductor-based digital signal detection and processing capabilities.[25]

The biggest beneficiary of these developments in the commercial sphere was two-way wireless service. This was made possible by the cellular network concept, invented at Bell Laboratories in the 1960s and 1970s.[26]

### Wireless voice goes digital

Cellular networks are a brilliant solution to a fundamental problem that had prevented wide deployment of wireless service. The previous two-way wireless architecture used scarce radio spectrum to cover wide geographic areas. This severely limited the number of users that could access the system at the same time.

The idea behind the cellular network is that spectrum can be reused by building adjacent regions, or cells, with their own transmitting and receiving towers, and limiting their range to that region, usually a few square kilometers. This allows adjacent cells to use the same frequency spectrum without interfering with each other.

As a wireless user moves from one cell to another, technology built into the system hands off connected calls to the appropriate

---

[25] These advances are fully described in T. S. Rappaport, *Wireless communications: Principles and practice*, 2nd edition (Upper Saddle River, NJ: Prentice Hall PTR, 2002).

[26] V. H. McDonald, "The cellular concept," *Bell System Technical Journal* 58 (1979), 15–43.

transmitter/receiver. The whole wireless network is connected to the wireline infrastructure so calls can be made to or from anywhere.

AT&T rolled out the first generation of cell phones, an analog system called Advanced Mobile Phone Service (AMPS), in Chicago in 1983. Once cellular systems entered the digital age with second-generation (2G) technology in the 1990s, progress was faster.

- The Global Service for Mobile (GSM) became the most widely used cellular technology in the world after its introduction in Europe in 1991.
- In North America and some parts of Asia another 2G technology, developed by Qualcomm and based on CDMA (Code Division Multiple Access), was introduced in the early 1990s.
- So-called 2.5G systems, launched in the late 1990s, marked the first packet-based cellular systems. They included GPRS (General Packet Radio Service), derived from GSM; and 1x EV-DO, derived from Qualcomm's CDMA technology.
- Now a 3G system, a packet-based technology introduced after 2000, offers much higher capacity, allowing data, voice, and video to be offered on handsets.

In the 1980s and early 1990s cellular phone service was still restricted to business use. It was too expensive for most individual consumers, even in developed countries.

Today basic service is affordable practically anywhere. For example, millions of low-income people in India and China have access to voice service because fees are only $8 to $10 a month, and handsets cost under $100.

By the start of 2006, only fifteen years after the introduction of the first digital system, there were more than two billion wireless users worldwide, including nearly 400 million in China alone.[27]

Cellular's amazing adoption rates were spurred by two factors. First, digital technology has created systems that make more efficient use of the available radio spectrum. Essentially this allows service providers to put more people on line. Second, there have been huge price cuts for both network equipment and user handsets, making the service far less expensive to offer and use.

Wireless communications was transformed into a global mass market by the cellular revolution of the mid-1990s. We are already seeing

---

[27] Morgan Stanley Research, April 16, 2006.

the emergence of the next stage: more advanced services, including data and video distribution to wireless handsets.

### Packets enable data transmission

Since digital wireless networks are packet-based just like the Internet, their messages can theoretically contain any kind of information. As a result, data transmission is increasingly moving to handheld wireless devices.

The first wireless data networks created dedicated connections that lasted the length of the data message, emulating the method used by PSTN data links. The newer approach uses encoded, switched data packets. These are transmitted on a channel shared with other messages, but the data packets are addressed to reach only a specific subscriber.

Switched-packet technology makes the most efficient use of available spectrum. It is the foundation of the new third-generation (3G) cellular systems. The most advanced wireless technology, under development for large-scale mobile markets, aims at data rates as high as 2 Mb/s, although in practice the rates are likely to be lower.

Wireless has also been used to connect fixed subscribers, but this approach has found a very limited market. The largest such networks (e.g., in Mexico) have reached only 500,000 subscribers. New standards such as WiMax promise lower costs and broadband capability to rival DSL. These might meet with greater success.

## Potential capacity limitations

The growth of IP-based networks has produced a classic case of demand rising in tandem with capacity. However, traditional telecommunications networks continue to be used, so big networks use mixed technologies. Voice, video, and data traffic are all increasing as companies and consumers alike call up, log in, and download. Even the phone companies are promoting TV and movie services over consumer digital subscriber lines. Is there enough capacity to accommodate this growth?

The answer depends on what kind of capacity we're talking about. It's unlikely we'll run out of data transport capacity in the backbone of the Internet, for example, because fiber optic links can be scaled to any capacity the routers can handle. Although the theoretical limit is much higher, in practice single fiber links will transmit 43 Gb/s over a distance of 4,800 kilometers without repeaters.[28]

Furthermore, a fiber can carry more than one channel of data streams by using different laser sources emitting at different closely spaced wavelengths (colors). Wave Division Multiplexed (WDM) systems use this technique to put as many as 160 different channels on a single fiber.

Finally, there is the matter of so-called "dark fiber." During the 1990s there was an extensive build-out of optical fiber cables to meet an anticipated surge in demand. The estimates turned out to be wildly optimistic. Anecdotal information suggests that only a few percent of the fiber links now available actually carry traffic, although the rapid rise in video transmission is likely to change this number.

In short, supply is still way ahead of demand. This has produced the expected result. As fiber capacity has increased, transport costs have dropped dramatically. An informal survey of large national companies that lease data transport to interconnect their facilities suggests that the monthly cost of a 100 Mb/s connection declined by 95 percent between 1995 and 2005.[29]

At the local level it's a different story. In this market segment data capacity is an issue, at least in the US, due to a combination of technical limitations and economic reality. Accordingly the cost of local data access at 1.54 Mb/s to a small business or residence has declined by only one-third, and even less in some regions.

Cost reductions in the local market haven't matched national figures for fiber mostly because the market is uncompetitive. Fiber has not been economical to deploy in local markets, so copper wirelines predominate. Wired "last mile" lines to consumers or small businesses are generally controlled by the incumbent phone or cable company, which is in a better position to control prices.

All this is beginning to change. As equipment costs decline, local fiber optic terminations will expand. Verizon Communications, one of the largest US telecommunications providers, has promised nearly

---

[28] Data quoted in Lefrancois, Charlet, and Big, "Impact of cumulated dispersion on performance," 174–176.
[29] Author's informal survey of major financial institutions in New York City area (2005).

universal broadband access in its regional market, and AT&T has a similar program under way.

For the foreseeable future, most small businesses and consumers will be stuck with copper lines or cable access with limited data carrying capacity. At the beginning of 2006, only about 32 percent of all US residences had what is called "broadband."

Since broadband in the US is presently limited to under 10 Mb/s, this is actually low-data-rate service by the highest world standards.[30] In countries such as South Korea, many people have much higher data-rate connections.

On the mobile side, wireless access rates are generally below 2 Mb/s, though fixed wireless access can reach much higher values.

## Software innovations

Computing hardware may be getting cheaper and more powerful every year, but without clever software it is useless. Software is the chief enabler of new technology-based products and services, including services based on IP networks. In addition, software development can represent as much as 70 percent of the development cost of new electronic equipment systems.

Figure 2.7 tracks US industry's investment in software over the ten years between 1993 and 2002. The increase from well under $100 billion to nearly $200 billion during that period underlines the importance of the field.[31] Given how much software is generated every year, it would be surprising if there had been no effort to systematize the writing of code.

And in fact, over the past fifty years the writing of software has gone from an arcane art to something more akin to a production process, employing millions of people.[32] This "mechanization" has been made possible by new methodologies, languages, and tools that automate much of the routine work originally needed to write and test software.

Although program definition and architecture remain highly specialized skills, the software development and maintenance process has

[30] Data from Cusick, "Thriving on telecom turmoil."
[31] SG Cowen Tech Research Team, Perspectives, "Tech road map 2002: Our outlook for the New Year" (New York: SG Cowen, January 2002), p. 105.
[32] D. M. Hoffman and D. M. Weiss (eds.), *Software fundamentals* (New York: Addison-Wesley, 2001).

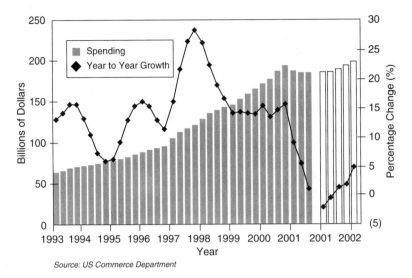

Source: US Commerce Department

2.7. Historical software investments in the US since 1993 (from SG Cowen, Perspectives, ref. 31).

been so streamlined that it can be outsourced to remote locations around the globe, where people without extensive education or experience write code and test programs. This became far more common as the industry shifted to object-oriented code.

Object-oriented programming is one of four innovations that underlie modern software technology. The others are relational databases, software as a service, and the open-source movement. Each has added new tools to the discipline. We will look at them in turn.

## Object-oriented programming

Of all the innovations that made the current advanced state of software technology possible, object-oriented architecture takes star billing.[33]

At its simplest level, a program is a series of instructions to a computer to perform a specific sequence of operations. A single programmer can track the logical flow and the desired outcome of smaller

---

[33] For a full explanation see G. Booch, *Object-oriented design with applications* (Redwood City, CA: The Benjamin/Cummings Publishing Company Inc., 1991).

programs. However, large programs involving complex interacting operations require a division of labor through which sequences of operations (called *subroutines*) can be defined and assigned to several different programmers. The subroutines are then assembled into the final program.

This seems simple, but in practice it has proven very difficult to manage. Programs evolve over time, with programmers working to improve performance or add features and functions. Even with the best of intentions, requirements defined at the start are guaranteed to change. If the program's architecture proves inflexible, changes become difficult, and the quality of the finished product suffers. Furthermore, because code from one program cannot be reused in another, each new program must be written from scratch.

Or at least that was the situation in the early days of software development. Object-oriented technology, a revolutionary idea conceived in the late 1960s and widely adopted in the late 1980s, provides a modular architecture that solves many of these problems.

**Objects and software structure**
Let's start with some definitions, and then look at what it all means. An *object* is a "bundle" of software that contains related data and the procedures, or methods, for using the data to perform specified operations. The methods are fixed, but the data can change as required.

Objects interact with each other by sending messages (via a bus) that request the execution of their individual procedures. A message consists of the name of the receiving object and the name of a procedure that the receiving object knows how to execute. The message may also include additional information (called parameters) required for a particular object to perform its procedures.

An object forms part of a *class* in which all the objects are identical in form and behavior but contain different data. Any object in a class is an *instance of the class*. The extraordinary value of the object concept is that it allows the formation of class hierarchies. This lets the software architect structure complex programs, from the most general down to the most specific elements, as a series of modules. Each module performs certain well-defined processing functions in the sequence designated by its place in the hierarchy.

Suppose, for example, that at some point the program needs digital signal processing functions for creating music, or a series of algorithms to solve problems in physics. The architect simply chooses classes of objects that carry out those procedures and plugs them into the program's bus. This is a much simpler and more logical way of representing complicated operations.

In the hands of skilled architects, any program can be structured as objects interconnected by a bus. The objects can then be coded individually in the expectation that the whole edifice will be brought together at the end as a functioning program.

Object-oriented technology has dramatically changed the way software is written.

- Once the architecture of a program is defined, development tasks can be divided into objects and each part independently tested.
- Once written, the objects are building blocks that can be reused in other projects.
- Changes in a completed program can be made by changing objects.
- Object-oriented technology produces more robust programs as:
  - each object hides the complexity of its function;
  - messages between objects convey the relevant variables needed for the object to do its job, but only the object's own methods interact with its data;
  - this prevents corruption of the object and, by extension, the program.

To see how object-oriented programs are actually built, turn to the example in Appendix 2.3.

## Relational database architecture

Managing the massive amounts of information used in a modern business requires a database, a highly efficient architecture for storing and retrieving content. The single most important innovation in that field is the relational database model

The concept is simple. The data is arranged in tables. New tables are created from older ones as data is presented or rearranged to meet new needs.[34] The term "relational" is a way of stating that tables are the basis of the system.

---

[34] For a complete description see C. J. Date, *An introduction to database systems*, 6th edition (New York: Addison-Wesley, 1995).

Introduced in the 1970s, relational database architecture has largely replaced the less flexible hierarchical model used in early mainframe computing. It dominates the field because its tabular formats store not only the information but the relationships within the information. Relational searches can be logically conducted with great efficiency, generating new tables to provide the requested information in a useful format.

The Oracle Corporation is now the world's leading supplier of proprietary software products for managing relational databases, and of tools for integrating them with other data processing applications.

## A new software model: Software as a Service (SaaS)

A convergence of innovations often produces unexpected outcomes. Software modules, now linked by the Internet, are again transforming the software industry by enabling what are generally called Web services or "software as a service" (SaaS).

In Web services, software residing on separate computers is assembled on demand for specific applications, then delivered to a user through the Internet. In effect, rather than having a software application residing on an individual computer, the user taps the Internet to get a software-based service.[35]

The delivery of software-enabled services to a user's terminal is not a new idea. In the 1960s the mainframe computer was connected to a "dumb" client terminal. In the client/server architecture of the 1980s, the server computer contained the business logic and data storage, while the client computer had the resources needed to run applications. In both cases the user could only call up applications that were available within the local area network.

In the new Web services model, however, the components needed to provide a given service are not resident on a single mainframe or server. They are distributed among multiple Internet-connected computers. This structure can deliver functionalities within and outside of an enterprise.

---

[35] For a full presentation see G. Glass, *Web services: Building blocks for distributed systems* (Upper Saddle River, NJ: Prentice Hall PTR, 2002); and A. T. Manes, *Web services, a manager's guide* (Boston: Addison-Wesley, 2003).

The SaaS Web services architecture represents a major step toward distributed computer networks. Linked by the Internet, these networks promise increased flexibility in the development and maintenance of applications software. They are capable of delivering services to or from any kind of device, from a mainframe computer to a small consumer gadget or a processor embedded in an instrument.

And since users can tap into applications through the Internet regardless of their location or computer platform, enterprises do not need multiple sites hosting the same software to achieve these benefits.

In addition, upgrades are no longer an issue. Everyone always uses the latest version of the software, since that's what's on the Internet.

While the potential benefits are easy to see, the implementation and maintenance of the architecture is not simple. In addition to developing frameworks for creating the services, providers must find reliable, automated methods for several important functions:

- registry (to capture the details of the services being made available);
- approval of work flow;
- policy management (to ensure that the services meet enterprise compliance standards);
- management of program component changes.

Because the architecture is so open, another key requirement is ensuring secure communications over the public Internet. Commercial SaaS products are designed to provide this capability by default.

There are many successful examples of SaaS implementations on a large scale, including the package-tracking system for Federal Express, which uses a BEA Systems platform. A simple example of a successful SaaS implementation is at Accredited Home Lenders, Inc., which is in the mortgage lending business. Accredited Home Lenders has reduced the procedures involved in completing a mortgage application to a set of interconnected software components. This has streamlined the process of managing applications by turning it into a simple, integrated, self-service procedure. Brokers go on line to submit applications, get pricing, and receive approval. In addition to saving time and money, the process ensures uniform procedures and consistent risk management.

The SaaS approach to deploying software applications is attracting a great deal of interest. However, it is not a trivial matter to master the library of functions and train developers to use the ones appropriate to a given task. Integrating activities in companies with multiple development sites significantly raises the stakes.[36]

In spite of the learning curve, though, the benefits of the new architecture are truly compelling. In mid-2005 about 70 percent of the largest companies were already using SaaS architectures for some applications. Most of the rest were in the process of developing applications of their own.

For these companies, the biggest hurdles in implementation are creating interfaces to existing legacy code, mastering security hurdles, and establishing standards.[37] Other issues include the challenges of routing packets that carry an enormous amount of information, some of it relating to security, and of maintaining routing service levels.

Bandwidth limitations and slow response can pose problems in some applications. Fortunately, as the processing power of chips increases, the speed with which packet analysis can be conducted improves as well. This will assure continuing progress in overcoming these limitations.

### Selling service. Not software licenses

You might wonder why all providers don't offer on-demand software functions instead of sticking to the model of loading large, full-featured stand-alone programs onto PCs.

Imagine not owning a word processor, office suite, or financial package. In their place you would have twenty-four-hour access to those programs and any others you might occasionally need. You would simply access the server or servers that house the required software and use it for the task at hand, sharing the application with others who are using it at the same time.

This sounds promising. But it does present several problems. The biggest of these is lack of control. Rather than having full possession of the application, the user is totally dependent on a lifeline to the Internet.

There are benefits to balance this disadvantage. The first, presumably, is lower costs. Companies would not have to license software

---

[36] See D. Karolak, *Global software development: Managing virtual teams and environments* (New York: Wiley-IEEE Computer Society Press, 1998) for a review of the challenges.

[37] Forrester Research Inc. (Cambridge, MA: April 2005); Evans Data Corp. (Santa Cruz, CA: June 2005) in *Computerworld* (August 15, 2005), 42.

packages in multiple copies, one for every user, knowing that they would sit idle most of the time.

They would not have to worry about their users working with outdated programs either. Instead of wasting users' time in occasionally updating PC-resident packages (assuming users bother to update them), they would be giving all users automatic access to the latest version of any program they need.

Solutions that use SaaS architecture are already available at the enterprise level. They are generally provided and maintained by corporate information technology organizations. But independent vendors are obviously anxious to offer them on their own.

The reason for their enthusiasm is easy to understand. The usual software business model, which involves selling software through licensing agreements and then negotiating maintenance contracts, provides a "lumpy" revenue flow. Software vendors prefer to provide software on a recurring "rental" basis, with customers paying a subscription. This generates a smoother, more predictable revenue stream.

The SaaS model has met with success, but problems have surfaced. One survey showed that 69 percent of the enterprises polled were satisfied with the architecture they had installed.[38] Of those that found the results disappointing, 50 percent found that legacy applications could not be integrated, 25 percent found that the cost was higher than in the previous architecture, and 25 percent found that the complexity and difficulty of maintaining the service was higher than expected.[39]

Problems have been most common in situations where the solution had to incorporate legacy applications. In brand new IT structures, on the other hand, the degree of success is limited only by the selection of software functions deployed and quality of the communications network.

Selling software as a service is a trend in tune with the times. Services have represented more than half of US economic output since 1987, reaching 82.5 percent of the total in 2005. According to a study by Peer

---

[38] C. Babcock, "Work in progress – service-oriented architectures are a blueprint for software development, but they're hard to perfect," www.informationweek.com (October 31, 2005) (accessed on April 20, 2006).

[39] www.itarchitect.com (November, 2005) (accessed on May 2, 2006).

Insight LLC,[40] service innovations are likely to be more successful if their development is centralized, which is certainly true of SaaS. However, they also require more attention to the customer experience than product innovations, as well as more prototyping. This may help explain the early problems in SaaS implementation.

## Computer operating systems and open-source software

Sparked by the open communications available through the Internet, the open-source software movement, which began as a student project, has revolutionized the software industry.

The name of the movement encapsulates its software development philosophy. Software developers make the underlying program code of open-source software freely available to everyone, to use and modify as they wish. Variations occur where such "free" software morphs into a licensable product as the result of specific modifications made by commercial companies.

The Linux operating system is the first and most important example of open-source software. Before we take a closer look at Linux, however, we should first clarify what an operating system is and why it is significant.

An operating system is a group of software routines that access the hardware of a computer and enable it to perform the functions requested by application software. Different computers have different operating systems, and application software must be tailored for the specific operating system of the computer on which it will run.

In the early years of the industry, a proliferation of computer manufacturers (most now defunct) generated an equal number of operating systems. After years of consolidation, however, Microsoft Windows and various versions of UNIX (an operating system that originated in Bell Laboratories), along with MVS, developed by IBM, emerged in the late 1990s as the dominant "proprietary" operating systems.

Obviously the ability to lock in applications programs that run on specific types of machines is a valuable asset for owners of operating

---

[40] J. M. Rae, "Using innovation to drive growth in service markets," presented at the Business Marketing Association annual conference, May 12, 2006. Thanks also to Ms. Rae, President and S. J. Ezell, Vice President, Peer Insight, Alexandria, VA, for providing a preliminary executive summary of their forthcoming study: *The discipline of service innovation*.

systems, and a good source of licensing revenues. This situation even offers users a benefit. Tight control over the core code of the operating system insures a desirable level of uniformity and interoperability.

However, no one willingly pays licensing fees if there is a viable free alternative. That is what Linux represents.

The Linux operating system was developed by Linus Torvalds while a student at the University of Helsinki in 1994. Intended as a simplified replacement for UNIX, it has had a dramatic impact worldwide on the cost of computers, and has acted as a stimulant to software innovation. The success of Linux is the result of a number of factors. Over the years, various computer manufacturers licensed UNIX but built in proprietary modifications in order to lock customers into their products. Hence computer buyers found themselves restricted in their ability to buy equipment from different vendors and felt that they were being overcharged. The advent of Linux opened their equipment options. Once IBM endorsed Linux, this operating system gained enormous commercial credibility.

**Free and valuable**
The success of open-source software caught everybody by surprise. Just imagine complex programs, even an operating system, for the most demanding equipment in the world, with no single point of ownership. Contrary to many prophecies, these products have achieved enterprise-level quality comparable to, or actually better than, long-established proprietary products.

Linux code is available free on the Internet, although few organizations can do without the support provided on a commercial basis by several companies. Linux has increasingly proven to be as robust as the UNIX operating system that it replaces. It has even been adopted by IBM.

No one owns Linux or any other open-source software. But open-source software does have control points. These are established by *ad hoc* communities of developers devoted to specific software sectors and working in concert.

The software is supported by organizations of volunteers who collaborate over the Web. They congregate at sites such as SourceForge.net, which calls itself "the world's largest Open Source software development web site, hosting more than 100,000 projects and over 1,000,000 registered users with a centralized resource for managing projects, issues, communications, and code."[41] Naturally, the number of projects that yield valuable software is but a small fraction of these endeavors.

There are thousands of groups that establish standards and decide which modifications will be accepted. For example, Linus Torvalds is the current authority for open-source Linux. Improvements that pass muster with Torvalds and the Linux standards committee are periodically wrapped into new Linux distributions that can be downloaded at no cost from the Internet. Companies such as Red Hat and Novell sell packaged versions with user support.

In contrast to proprietary software, which is relatively static and "closed," open-source software evolves constantly. Useful changes and corrections are made available very quickly and posted on the Web for all to see and use.

Open-source software engages the coordinated interest of many developers with no axes to grind except the satisfaction of seeing their work accepted by others. Their efforts drive the fast turnaround of new features, functions, and fixes.

But that does not mean that the software is slapdash. Release cycles are managed by the volunteer organizations to ensure quality and provide rapid fixes for any deficiencies. The end product is often touted as being comparable to proprietary software programs.

### Quality, performance, competition

Questions regarding open-source, mission-critical software usually revolve around issues of quality control, network operability, and security features. Over time these concerns have abated, thanks to the speed with which patches and vulnerabilities are identified by the open community and posted on the Internet for all to see and react.

One big business that will be increasingly impacted by the availability of dependable open-source software is the database sector. The field has long been dominated by Oracle. Now, however, there's an open-source competitor, MySQL, which claims to offer an alternative free of licensing fees. It comes with impressive credentials. The Los Alamos National Laboratory has used it for several years with good success.[42]

---

[41] http://sourceforge.net/docs/about (accessed May 23, 2006).
[42] M. DiGiacomo, "MySQL: Lessons learned on a digital library," *IEEE Software* 22 (2005), 10–14.

MySQL is not alone in claiming parity with commercial products. A number of studies[43] have been made regarding the quality standards of open-source software for mission-critical applications, and the results have been quite satisfactory.

The fact that open-source software is license-free has not escaped the attention of governments. Government agencies looking for a no-cost alternative to Windows and UNIX helped foster the international acceptance of Linux. Authorities in China, Taiwan, Hong Kong, and South Korea, as well as countries in Europe and Latin America, have made it a priority to adopt open-source software whenever possible. Taiwan is particularly important, since so many of the world's computer boards are produced there.[44]

Open-source has defied the doubters who considered it a hobbyist enterprise. It is estimated, for example, that some form of Linux was used by 80 percent of major companies in 2005.[45] Thanks to the success of Linux, the open-source idea has gone from an experiment to a very practical way to save money while encouraging innovation.[46]

With all that is at stake for commercial companies, it is no surprise that litigation plagues the open-source industry. Claims have been made that parts of the Linux operating system are elements of the copyrighted UNIX code base. Even users are at risk. Because licensed and unlicensed software is commonly used in the same environment, they must make sure that the licensable software is accounted for and licensing fees paid.

### An experiment in success

The idea that a self-generated community of developers using the Web can develop complex commercial-grade software is truly revolutionary. While the number of successful projects is small relative to the number started, open-source activity will no doubt be the impetus for

---

[43] See, for example, J. S. Norris, "Mission-critical development with open-source software: Lessons learned," *IEEE Software* 21 (2004), 42–49; T. Dinh-Trong and J. M. Bieman, "The free BSD project, a replication case study of open source development," *IEEE Transactions on Software Engineering* 31 (2005), 481–494.

[44] Opinion by M. Betts, "Linux goes global," *Computerworld* (July 18, 2005), 31–40.

[45] D. Schweitzer, "Linux muscles into Microsoft space," *Computerworld* (May 30, 2005), 33.

[46] F. Hayes, "Our Linux values," *Computerworld* (July 18, 2005), 46.

even more important innovations. In effect, developers are free to invent software, share it freely with all comers, and, once a community of users is established, have an informal arrangement for its further evolution.

I have seen the advantages of open-source development first hand. One of my portfolio companies used the approach for one application in a communications system.

The whole solution, including protocols, was assembled in about one man-month of development time. No licensing costs were incurred. In the past such an application might have required ten times the programmer effort and thousands of dollars of specialized software licensing fees.

An interesting side benefit of the open-source movement is the speed with which the market delivers its verdict on the quality of specific applications. As an ever-growing number of potential users gets used to looking for useful free software, all open-source programs are subjected to a very public popularity test.

Recently there have been some interesting open-source-related developments on the commercial front as well. One trend has a few established companies weighing the idea of opening up their proprietary products to public scrutiny.

One of the biggest is SAP, the world leader in enterprise resource software. In 2003, the company decided to make its software more flexible. It plans to accomplish this by reengineering its products to take full advantage of object-oriented technology linked with Internet connectivity. The task is massive, and will cost billions.

At its conclusion, SAP will offer customers a new architecture in which software systems consist of connected applications running on different computers linked through the Internet. This will replace its currently closed architecture.

This ambitious, far-reaching initiative will take years to have a significant impact on the ability of customers to allow more efficient sharing of data and resources.[47]

The second development moves in the opposite direction, from open to closed. New companies are being launched to exploit free open-source

[47] A. Reinhardt, "Germany: SAP: A sea change in software: The German legend's move into Lego-like modules could revolutionize the way companies do business," *BusinessWeek* (July 11, 2005), 46–47.

software as a starting point for their own commercial products. These products are designed to compete with offerings from established proprietary software vendors. However, their economic success is unproven because innovation inevitably produces a proprietary element in software that the market needs to validate.

For example, a new company called JBoss, Inc. is competing with IBM's WebSphere on the basis of open-source software. While these products address a market estimated to be in excess of one billion dollars, their success in garnering a meaningful percentage is unclear because established large software vendors are not sitting idle in the face of new competitors and continue to build increased value in their proprietary products.

Like other innovations in information technology, open-source software has created opportunities for new service businesses. A number of companies such as Red Hat have been established to provide support services for open-source code. By helping users implement, run, and customize the software, they offer the kind of comfort level that would have been provided by vendors of proprietary software.

While open-source software is free of license fees, many users wind up paying for this third-party support. Even taking the fees for such services into account, however, the overall cost of open-source solutions is still substantially lower than that of proprietary software.

## The digital future is now

Clearly, object-oriented technology, software as a service, and open-source software have changed our assumptions about the way software is developed and sold. At the same time, they are heavily dependent on the availability of networking and communications, especially the Internet. Communications, in turn, relies on the speed and power of computers.

In the rest of this book we shall see what impact all this technological innovation has produced on financial investment, corporate strategy, manufacturing, global employment patterns, and other non-technical matters. Every advance we have covered up to this point, from devices to systems to software, has resonance beyond electronics, and is forcing change around the world in the broader spheres of social and economic development and government policy.

# PART II

# Innovators, entrepreneurs, and venture capitalists

# 3 | *Edison's legacy: Industrial R&D*

> What makes the capitalist system work,
> what keeps economies dynamic, is precisely
> nonconformity, the new, the unusual, the
> eccentric, the egregious, the innovative,
> springing from the inexhaustible inventive-
> ness of human nature.[1]

Most of the revolutionary innovations in semiconductors and software explored in the preceding two chapters came out of the corporate R&D organizations of AT&T, General Electric, IBM, Corning, RCA, and other large companies.

That doesn't mean these breakthroughs reflect a corporate mindset. Nothing could be further from the truth. The scientists who created them were following their own paths, not some managerial directive or mandated marketing strategy. As someone who began his career in this environment, I can attest to its creative ferment.

Truly creative people tend to be free spirits. They are the essential spark plugs of innovation. At the same time, their drive for discovery is often in conflict with practical considerations such as company direction or fiscal constraints. This trait presents corporate R&D managers with endless challenges.

Perhaps that is why the talk today is all about "small, innovative companies," not big, powerful corporate laboratories (labs). But the fact is that for much of the twentieth century those big labs were the primary generators of new technology, because they had the resources and the funding necessary for research. They were the vehicles innovators needed to carry their work forward.

---

[1] P. Johnson, "What Europe really needs," *The New York Times* (June 17, 2005), A14.

We've begun our study of digital electronics by looking at how the technology works, and how it affects people and businesses. This chapter traces the history of the central corporate laboratory as a model for innovation, starting with its great progenitor Thomas Alva Edison.

It also looks at how market pressures and new technology eventually relegated advanced corporate R&D to a secondary role in exploiting the massive new business opportunities created by digital technologies. This marked a transition to widespread entrepreneurship backed by venture funding, the subject of a future chapter.

## Learning the ABCs

Before we discuss research and development centers, it's important to understand the three distinct kinds of R&D that were done at the corporate labs. Here are some generally accepted definitions.

*Basic research* aims at the discovery of laws governing natural phenomena. The best basic research is conducted in an open environment where creative minds can follow their instincts.

Major discoveries are frequently made in the course of looking for something else, so it's important that a researcher is free to work on hunches. From my experience, management-directed basic research is rarely productive, because basic research scientists are motivated by the excitement of exploring the unknown and earning the respect of their peers.[2]

Investing in basic research is a high-risk endeavor. It has the potential to generate enormous economic gains, but it is impossible to predict whether a specific discovery will have commercial value.

Adding to the risk is the need to invest in follow-up development before a raw technology can be commercialized. This investment is costly, since development work is best done with the collaboration of scientists who are expert in the subject. Expertise of this kind does not come cheap. Even rarer is the expertise to exploit the new market opportunities created by major innovations.

The fundamental work that uncovered the properties of semiconductors in the 1930s and 1940s is an excellent example of the process. As basic research, its discoveries produced extraordinary

---

[2] An interesting perspective on this subject is found in B. Bunch, *The history of science and technology* (New York: Houghton Mifflin Company, 2004).

technological and commercial results. But these outcomes required extensive applied research and product development before they could fulfill their potential.

*Applied research* occurs farther along the road to productization. It begins with a general knowledge of basic physical phenomena and seeks to discover a path to their practical utilization.

Applied research can have far-reaching consequences. The Bell Labs work that made it possible to prepare single-crystal germanium and establish its properties led directly to the construction of the first transistor. Early work on data structures was the stimulus for the relational architecture underlying modern database technology.

*Product development*, the third type of research, covers all activities that lead to new products or processes, help make existing products better, or improve their manufacturability. In the case of transistors, for example, product development work was focused on finding ways to fabricate practical devices.

There is no question that, in the long term, innovation is impossible without basic research. Scientists have to understand fundamental principles and materials before they can apply them to new products. But once this understanding is in place, the primary activities of innovation are applied research and product development.

Dr. Satyam Cherukuri provided an elegant illustration (Figure 3.1) that helps explain the product innovation process.[3]

The S curve on the chart tracks the stages of innovation over time, from concept to product maturity. Innovation moves through stages labeled A, B, and C.

A. Innovation starts with the three phases of the *product development* stage.
  - The *concept* phase is where product ideas are generated. These concepts usually grow out of applied research.
  - In the *feasibility* phase, the ideas are refined to the point where it is possible to estimate a product's economic value.
  - Finally, during the *prototype* phase, a working model or service trial demonstrates the basis for a commercial product.

---

[3] Figure adapted from presentation by Dr. S. Cherukuri, CEO of Sarnoff Corporation (2004).

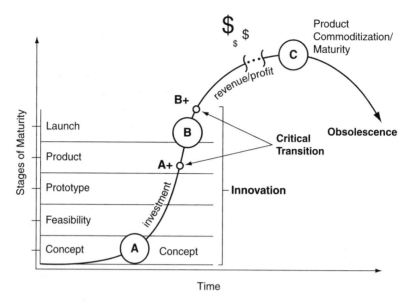

**3.1.** Dr. S. Cherukuri's curve of product maturity over time showing the process from concept to obsolescence. Reprinted with permission from S. Cherukuri, Sarnoff (ref. 3).

B. The process now proceeds to the critical (and costlier) *product launch* stage. If the launch is successful, this stage delivers the high profit margins and return on investment necessary to encourage continued innovation.

C. Eventually the product enters the *commoditization* stage. Competition for the product emerges, driving profit margins down. When obsolescence sets in, the product loses all commercial value.

The path from A to C can span decades. In the case of the transistor, five years elapsed between the first demonstration in 1947 and the first commercial product. On the other hand, in some consumer electronics sectors, the time from idea to product launch can be as short as two years.

By clarifying the process of product innovation, Dr. Cherukuri's chart also shows where the hazards lie. Right at the start, for example, a misreading of the market or of the competition can produce a fatally flawed product concept. This misstep only becomes apparent during the launch, when the finished product fails to meet sales expectations. By that time the producer has already invested heavily in its development.

Failures at other points in the process also cause products to under-perform. A product may prove too difficult or expensive to manu-facture. If development takes longer than projected, competitors can stake out the market with innovative products of their own.

So much depends on the innovation process that it's not surprising that companies try to manage it. In this field, as in so many others, the person who created the pattern for the modern era was its greatest innovator, Thomas Edison.

## Edison invents the industrial laboratory

The twentieth century has often been called the Age of Edison, because he played a role in creating many of the technologies that shaped the modern industrial world. With 1,093 patents to his name, Edison has been held up as the archetype of the lone inventor, producing inno-vations by sheer force of personal genius.

This is a myth. Thomas Edison did not work alone. He may have been the driving force behind his many inventions, but he was sup-ported by a hand-picked research team assembled for that purpose.

The composition of his team marked a change in attitude toward scientific endeavor. Since the beginning of the Industrial Revolution there had been a sharp cultural distinction between "pure" scientists who advanced the state of knowledge, and practical inventors of com-mercial products. Scientists enjoyed high status as noble searchers after Truth, while inventors were considered a lower order of commercially-minded technicians. The scientists did basic research, while the tech-nologists did applied research and product development.

Recognizing that this cultural gap must be bridged within a single organization is one of Edison's key contributions. In 1876 he created the world's first industrial laboratory in Menlo Park, New Jersey. This organization, which he dubbed the Invention Factory, successfully integrated scientists with technologists. His move to the Edison Laboratory in West Orange, New Jersey in 1887 solidified his organi-zation and created what has been called the world's first R&D center.

The spectacular results from his labs made history, producing the phonograph, motion pictures, incandescent lighting, and electric power generation (among many other breakthroughs). His discovery of the "Edison effect" in 1883 anticipates modern electronics, even though he himself did not pursue it.

The phenomenon he described, of electrons flowing to a metal plate placed in one of his light bulbs, inspired John Ambrose Fleming to create the world's first diode in vacuum tube form. This led, in turn, to the creation of the triode, a tube capable of amplification and counting, by Lee deForest, an early radio pioneer.

Edison's labs generated a steady flow of products. What's even more impressive is how many of them were commercially successful. This was not an accident. Edison designed his approach for that outcome, based on hard experience.

In 1868 he had patented his first invention, an electric voting machine. The idea was to speed up the legislative process by eliminating time-consuming roll calls through instant tallying of yea and nay votes in the Massachusetts house (he was living in Boston) and in the US Congress.

The machine worked perfectly, but found no buyers. Legislators liked roll calls because they represented an opportunity for oratory. The lawmakers also used the delays during roll calls to cut deals with other lawmakers for the passage or defeat of the bills being debated.

After this disappointment, Edison decided he would never again work on a technology that didn't have a good chance of commercial success. Research for its own sake played no role in his investigations. All R&D was focused on the development of product innovations – the A and B stages of Figure 3.1 above.

If the research lab is Edison's greatest invention, as some have claimed, his focus on teamwork and creating products makes his lab the forerunner of all commercial research strategies. One perceptive writer has summarized his significance.

A just view of Edison's historic role, suggested recently by Norbert Wiener, would be that he was a transitional figure in late nineteenth century science, arriving on the scene at the period when the crude mechanical inventors had done their part, and systematic experiment and research was henceforth to be undertaken by skilled and specialized men, on a much larger scale than before ... Today [1959], giant research laboratories, owned by private corporations or by government, stand everywhere as monuments to Edison's innovative spirit. The technicians working in those institutions have been telling us for some time now that the day of the lone inventor, in an attic or small laboratory, has passed forever. Inventions, we have been assured, are henceforth to be produced by scientific teams using more complex instruments than Edison ever knew and "made to order." But can we ever dispense with the individual inventor, who is both dreamer and man of action? Good inventions are still not "predictable." Of late, serious doubts have been expressed that the mile-long laboratories and the teams of experts are any guarantee of original achievement. Technical men themselves voice fears that the new conditions of mass research may be less helpful in independent or nonconformist thought; and that the bureaucratic inventors of today may be losing something of the intuitive skills and the sense of simple things that an Edison possessed.[4]

As countless companies have since learned, managing creative industrial laboratories is difficult. Unfortunately, Edison did not leave behind a foolproof recipe. Here is a description of a visit to the Edison Laboratory in 1900, when its staff had reached 90 people:

A sociologist wandered into the Edison Laboratory one day to inquire of the great inventor what methods of organization he used in conducting his research work. "I'm the organization," Edison said roundly. The professor could discover that for himself by looking around for a few minutes, and could save himself much time thereby. In truth the sociologist soon learned that, though the inventor had numerous assistants and business managers, he kept control of everything in his own hands, and himself supervised all the different experiments and tests going on. Yet somehow the "system" worked.[5]

## The big central research laboratories emerge

Edison's example led to the formation of several large corporate laboratories in the twentieth century. The intent was to provide an environment for far-reaching innovation which would be commercialized by product divisions. In order to eliminate short-term product pressures on the labs, funding for their activities came from the corporate budget rather than budgets tied to specific product units.

In terms of Figure 3.1, the labs were expected to participate in the A to A+ stages, leaving B to the product divisions. Of course, implicit in this arrangement is the corporate willingness to maintain predictable

---

[4] M. Josephson, *Edison, a biography* (New York: History Book Club, 1959), p. xiv.
[5] Josephson, *Edison*, p. 411.

and constant annual funding, regardless of the vagaries of business cycles.

The strategy also assumed that the product divisions would have the ability to exploit the innovations developed in these laboratories for commercial gain. Obviously, it was not expected that these laboratories would be free from the necessity of integrating their results with corporate strategic needs.

The most important electronics research laboratories were part of vertically integrated corporations such as American Telephone and Telegraph (AT&T), International Business Machines (IBM), the Radio Corporation of America (RCA), Westinghouse, and General Electric, covering broad industrial sectors.

The biggest of them all was AT&T's Bell Laboratories, founded in 1924. The work at Bell Labs ran the gamut from basic research through product development. By the early 1980s, when AT&T was broken up by agreement between AT&T and the US Department of Justice, Bell Labs employed about 20,000 scientists, engineers, and support staff distributed in various facilities around the country. A total of 2,000 were in the research department, while about 8,000 people worked on the software that operated the telephone network and enabled its services. The remaining 10,000 were tasked with product development on behalf of AT&T's business units.

Long-term research and product development spanning many years were possible at the Laboratories because they benefited from a unique industrial advantage: being part of a government-created telephony monopoly.

AT&T controlled the telecommunications network, its technology, the production of all telephone equipment, and (through the regional Bell operating companies) most service deployment. AT&T's profits were predictable, because they were guaranteed by cost plus pricing (based on assets deployed) negotiated with government regulators. As a result, the rate at which new services and technologies were introduced was controlled by the company – and because it had no competition, it could take its time.

This happy situation made it possible for AT&T to support a very large R&D staff covering all aspects of science and technology with a bearing on communications, and to provide a well-planned path from research results into products. Competitive technologies could not threaten the business, because they were not allowed. This was one reason why packet-switched networks did not get serious attention from AT&T when Internet Protocol-based technologies emerged in the 1970s.

Society benefited from this structure, too, at least through the 1970s. Bell Labs work that spanned decades made possible the modern wireline and wireless telephony systems, and built the scientific and technological foundation for the semiconductor industry. The Laboratories also made noteworthy contributions to computer science, including the invention of UNIX, the most important operating system for large-scale and high-performance computing.

There is a common misconception that the big laboratories were unproductive ivory towers, where engineers and scientists worked in isolation. The example of Bell Labs proves that this is not true. While some of the basic research conducted by these organizations was esoteric, the vast majority of the research programs were in areas with commercial potential.

Many of these programs did not pan out, but that is the nature of the innovation process. In fact, only a fraction of the staff at these labs (typically about 10 percent) worked on research projects that were not tied into ongoing product development.

Because of the variety of skills among the researchers, the availability of first-rate facilities, and the ability to pursue programs that took years to complete, many of the great innovations that have shaped the modern world originated as a result of the work of a relatively small number of very creative people in Bell Labs and other laboratories. These are just a few prominent examples:

- IBM's research laboratories were early pioneers in electro-optic devices as well as mass storage technologies.
- Westinghouse Research Laboratories contributed active matrix liquid crystal displays as well as early research on integrated circuits.
- Hughes Research Laboratories developed the first practical gas laser.
- The Xerox Palo Alto Research Center developed the visual computer interface, the Ethernet protocol and the laser printer.

Another major contributor, RCA Laboratories, where I did my scientific work, will be discussed below.

Yet no matter how creative these organizations were, their value to their parent companies was usually beyond the control of their management or staff. They were sometimes the objects of criticism, mostly because of their companies' inability to commercialize their major innovations.

Here is where corporate leadership must play a key role.

### Great innovations need visionary leaders

"I'm the organization," Edison said. Not every big corporate lab had an Edison to run it. This is the story of one that did.

While relatively small compared to the other major corporate labs, RCA Laboratories in Princeton, New Jersey, was unique in its corporate leadership and the level of support it received from RCA. This made it possible for the Labs to generate a stream of outstanding commercial contributions far out of proportion to its size, including color television, reliable semiconductor laser diodes for telecommunications and other applications, the LCD, amorphous silicon solar cells, and many more.

The technical staff at RCA Labs was relatively constant for many years. It consisted of about 1,200 scientists, engineers, and technicians, including about 300 PhDs in disciplines ranging from basic physics and chemistry to electronic systems.

It was fortunate in its parentage. RCA was one of the pioneering companies in electronics. It was founded in 1919, initially to commercialize radio technology, and by World War II had become a leader in consumer electronics.

During much of its history RCA was under the leadership of David Sarnoff, a visionary who focused on ensuring the company's future through heavy investments in innovative technologies. Unlike Edison, Sarnoff was a businessman, not a technologist, but he had faith in the potential of electronic technology to transform society and create whole new industries. This was matched by a sure instinct for the market potential of individual innovations.

To ensure a steady flow of marketable innovations, Sarnoff decided to amalgamate earlier RCA development laboratories into a central corporate research facility. This led to the founding of RCA Laboratories in 1941. During World War II, the Labs focused on military communications, but after the war, commercial communications, consumer electronics, and particularly television became its prime areas of activity.

The development of color television, the achievement that made the Labs famous, is a great example of how Sarnoff's visionary leadership created an industry by linking brilliant research to a sound commercial strategy.

After the success of commercial radio in the 1920s, Sarnoff saw huge opportunities in television. RCA had demonstrated commercial black and white television at the 1939 New York World's Fair, but the war put a halt to further development. After the war, Sarnoff resumed his focus on television, with color television as the ultimate objective.

Color TV presented enormous technical problems for the technology of the day. Sarnoff turned to the Labs to solve them. He provided unstinting support in the effort, including spending the (then) enormous sum of $40 million on this program. He was roundly criticized by his many public detractors for squandering the company's resources on an innovation they thought the public wouldn't buy, and one they were sure couldn't be done.

None of this mattered. David Sarnoff could safely ignore his critics, because he had absolute control of RCA. And since RCA also owned the National Broadcasting Company (NBC), he was in a position not only to pioneer color television technology, but to ensure its commercial launch.

Urged on by Sarnoff, RCA Laboratories researchers invented and patented every one of the core technologies (including manufacturing methods) for color television by 1950. NBC launched the new service in 1951.

Its ultimate success was due to a sound technical strategy. It was designed to be compatible with existing television receivers. Color broadcasts could be viewed on black and white sets, while new RCA color-enabled receivers displayed both black and white and color broadcasts. This meant that color TV broadcasts could be initiated without losing viewers.

To create the equipment industry (and incidentally gain an audience for NBC), RCA also adopted an aggressive (and lucrative) technology licensing strategy that made it possible for other manufacturers and broadcasters to enter the color television market.

The development and successful commercialization of color television by RCA is an excellent example of the value of truly innovative research in a vertically integrated company driven by a visionary leader. No independent organization could have succeeded without RCA's access to capital, its broad technical and scientific skills, and its control of a television broadcast network. The strategy of strong intellectual property protection resulted in an impregnable patent portfolio that allowed RCA to dominate the television industry and extract license revenues from all parts of the world.

This is not to imply that the profits came quickly or easily. In fact, it took ten years for RCA to begin earning a profit on its color TV receivers. But the wait was worth it. Because of the profitability of its television products, RCA was able to expand into many other commercial and government product sectors.

Since it was a leader in vacuum tube electronics, semiconductor devices were a natural and important investment area for the company. The first germanium transistor designed for consumer electronics was developed at the Laboratories in 1952, and manufactured by the newly established Solid State Division in 1953. This was the start of a new era for the company, and for the electronics industry as a whole.

Not many companies get the opportunity to create a new industry. Even fewer meet the challenge. RCA did so, and changed the world forever. Most aspects of life today are in some way touched by television. The technology provides news and entertainment that would have been unobtainable just 60 years ago. It spawned a major consumer electronics industry, which has made it possible to provide a television set for one out of every four people in the world – a total of nearly two billion TVs in operation today.

Part of the credit for this immense achievement must go to David Sarnoff and the management of RCA. But an equal share should be assigned to the scientists and managers of RCA Labs. Like Edison, they proved that an R&D center can successfully generate revolutionary innovations as part of a corporate strategy.[6]

## The role of innovation management

An important element in the success of RCA Laboratories was a management style and culture that merged individual initiative and corporate product requirements. As a researcher and manager at the Labs for nearly twenty years, I had the opportunity to observe and work within this unique culture myself.

My experience convinced me that technology transfer into products is best done when innovators can participate in the process as much as possible. A "hands-off" approach is not as efficient, because an innovator who knows the product from the ground up can solve

---

[6] For a pictorial history of the Labs, see A. Magoun, *David Sarnoff Research Center – RCA Labs to Sarnoff Corporation* (Charleston, SC: Arcadia, 2003). A good biography of RCA's legendary leader can be found in E. Lyons, *David Sarnoff* (New York: Harper & Row, 1966).

performance and production problems far more rapidly than someone uninvolved with its creation.

That's why the Laboratories encouraged researchers to participate in moving their innovations into products. As part of this culture, PhDs willingly worked with engineers in the product division to launch products. Researchers commonly took up residence in RCA factories in the course of new product introductions. When the product launch was completed, they went back to the Laboratories and started other projects, applying the valuable practical experience they had gained, plus an appreciation of what it takes to deliver a product.

### Personal exposure

To promote creativity among new researchers, PhDs entering the Labs were encouraged to define their own areas of interest. I decided to study gallium arsenide (GaAs), then a little-understood semiconductor which had been successfully used to produce light emitting devices.

As a result, I met Dr. George Dousmanis, one of the scientists studying semiconductor lasers made of GaAs. He demonstrated the operation of these devices to me. At that time they were practically useless, because they operated for only a few minutes before burning out for reasons unknown.

Dr. Dousmanis, working with Herbert Nelson, enjoyed studying the physics of these new devices, but shared the general opinion that they were fated to remain laboratory curiosities. In fact, the funding for the laser program at the Labs came from the US Army, which was interested in developing a solid-state infrared searchlight for battlefield illumination.

To everyone's shock, Dr. Dousmanis died suddenly a few days after our meeting, leaving no one to complete his contract. I volunteered to take on the project, and was fortunate enough to discover the factors affecting the performance of these lasers. In addition, I conceived a new heterojunction architecture for the devices and developed the manufacturing technology, which was transferred to the Solid State Division.

In just two years this revolutionary device went from laboratory curiosity to the first commercial semiconductor laser. It was made commercially available by RCA in 1969 and found immediate applications in military systems.

This rapid effort was possible because the Labs' organization was so flexible that talent needed for specific programs could be assembled very quickly. The same story was repeated many times in the history of RCA Labs.

**Innovations for the core business**

Television products remained huge profit generators for RCA, but thanks largely to the work done at the Labs, the company successfully entered into other major consumer, military, and industrial markets starting in the early 1950s. During the 1960s and 1970s RCA Labs (renamed the David Sarnoff Research Center in 1951) pioneered such important innovations as:

- novel high-power silicon transistors;
- thyristors and rectifiers;
- the first metal oxide semiconductor field-effect transistors (MOSFETs);
- the first complementary metal oxide semiconductor (CMOS) devices;
- the first practical liquid crystal displays;
- the first practical solar cells;
- the first practical semiconductor laser diodes using heterojunction structures;
- the first rewritable optical storage disk;
- the first magnetic core memory;
- a host of new microwave devices.

Not all of the work was R&D. When other organizations within RCA did not have the talent they needed, the Laboratories stepped in to complete practical systems and products. Two examples warrant special mention.

First, the Laboratories assumed responsibility for providing the communications equipment that connected the Lunar Orbiter with the Lunar Explorer Module on the moon during the 1969 Apollo mission. I developed a core semiconductor device for this system.

Also, the Labs developed the first commercial-quality portable color TV camera to be built with CCD solid-state imagers (see Chapter 1). It was initially used by NBC – another example of the value of being part of a vertically integrated company. The achievement won the Labs an Emmy Award in 1984.

RCA Laboratories was a challenging place to work. But it was also the best place to be if you wanted to pursue interesting innovations, and then see them launched in product form.

## The 1970s: Corporate labs in a changing environment

After World War II, with the industrial base of both Europe and Japan largely devastated, the US came to dominate the technology-based industries of the world. New technologies in electronics and other fields, coupled with a manufacturing capacity untouched by war, gave US companies a huge advantage over their foreign rivals.

This situation would not last. In Europe the Marshall Plan jump-started post-war recovery. On the other side of the world, the enormous effort to rebuild the industrial base of Japan, starting in the 1950s, soon produced a formidable competitor. Japan's approach involved close collaboration between government and industry, as has been well documented.[7] It also involved licensing foreign technologies for domestic manufacturing.

The full force of competition started to hit American corporations in the 1970s, led by Japanese companies that had organized themselves to achieve leadership in selected industries with the collaboration of their government.

Figure 3.2 shows the US trade balance in electronics sectors from 1970 to 1989. This includes data processing equipment, electronic components, and telecommunications equipment. Note the decline in the US trade balance and the corresponding rise in that of Japan and other Asian countries.[8]

Growing international competition impacted various US companies differently in the 1970s, but in every case it had a negative effect on product profitability. This forced companies to revise their business strategies to counter the onslaught. The companies had to mobilize all of their resources to meet the challenge.

Their reorganization spelled trouble for many R&D labs. Companies came to the conclusion that they could no longer afford to finance the long-range research programs that their labs had carried out. They needed new products, and they needed them fast.

To force the central labs to focus on shorter-term projects, they were increasingly made responsible to corporate business units. The labs served these units as internal contractors, with their funding tied to specific projects.

---

[7] See C. Johnson, *MITI and the Japanese miracle* (Stanford, CA: Stanford University Press; 1982); and P. Choate, *Hot property, the stealing of ideas in an age of globalization* (New York: Alfred A. Knopf, 2005).

[8] P. Guerrieri, "Technological and trade competition: The changing positions of the United States, Japan and Germany," in *Linking trade and technology policies*, M. Harris and G. Moore, (eds.) (Washington, DC: National Academies Press, 1992), p. 44.

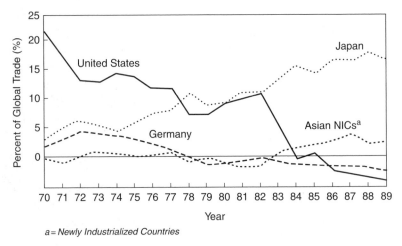

a = Newly Industrialized Countries

**3.2.** Trade balance in R&D-intensive electronics sectors, 1970–1989. This group of sectors includes data processing equipment, electronic components, and telecommunications equipment. Standardized trade balances are expressed as a percentage of total world trade in R&D-intensive electronics sectors. For methods and sources, see ref. 8. Reprinted with permission from *Linking trade and technology policies*. Copyright © 1992 by the National Academy of Sciences, courtesy of the National Academies Press, Washington, DC (ref. 8).

This represented a fundamental shift from the days when R&D efforts were totally funded by the corporation. Now, instead of annual funding from the corporate budget, R&D labs typically received only 10 percent of their operating expenses from headquarters. The rest was to be funded by government contracts or by direct allocations from business units in support of dedicated projects. These contracts were negotiated each year, and it is not hard to imagine why business units with low profits or outright losses were not very generous.

In effect the corporate labs gradually lost the status of "independent" research organizations. As budgets shrank, staff members faced the choice of selling their services to business units, competing for government contracts, transferring to the engineering departments of the business units, or leaving the company.

Many researchers, unwilling to surrender more of their freedom in selecting projects, left the companies to join the faculties of universities or work in government laboratories. This migration generated major new centers of innovation, as we discuss below.

At RCA Labs, too, an era was ending. Corporate funding was becoming an annual problem. Government work (mostly through the Department of Defense) was supporting a significant portion of the work on lasers, integrated circuits, microwave devices, and such other electronic innovations as liquid crystals and solar energy converters. Many of these eventually found their way into the commercial world, either under the RCA name or as products made by others under license from RCA.

Licensing, at least at RCA, was proving to be a lucrative business. Ironically, while RCA and many other companies were reducing their commitment to R&D, RCA's own experience was showing that research could add significant profits to the bottom line. When General Electric acquired the company in 1986, its annual license revenues of $160 million from over 28,000 patents greatly exceeded the $80 million operating budget of the Laboratories.

## The decline of the central laboratory

Even at the height of their reputations the central laboratories did not lack for critics. But once the business units took over their funding they were increasingly regarded as expensive luxuries with only tenuous connections to the businesses they served. The complaints against them included irrelevance, poor response time, and waste.

The problem stemmed from a mismatch between the way the laboratories operated and the new corporate imperatives. For example, the labs encouraged broad creative thinking as a way to stimulate researchers to do their best work. But too often this produced innovative product ideas that did not fit within the defined parameters of the company's business units.

It was easy to forget that the charter of such laboratories was in fact to open new market opportunities. Generating promising product ideas was a good beginning, but ultimately entering a whole new business involves investment decisions that only CEOs can make. More often than not, the "prudent" decision is to forego such new ventures and focus on the current business as part of the drive to maximize profit and minimize risk.

Faced with corporate indifference to the market potential of their work, more of the creative people left, taking their ideas with them. Many of their ideas later reappeared in the guise of new businesses funded by venture capital.

Another charge against the corporate labs involved their performance in product development. Critics complained that even when they came up with really promising products that everyone agreed were worthy of bringing to market, corporate laboratories were far slower in handing them off to production than the new nimble competition, foreign or domestic.

Paradoxically, one of the strengths of the central laboratories was also becoming a liability for their corporate parents. The expertise they had built up over many years in selected areas of technology was not necessarily what the corporation needed to satisfy new market demands. Too often the laboratories were unable to respond to challenges from new competitors who were using other, more efficient approaches. In short, their capabilities were out of date, and the experts they had nurtured for years could not easily shift into new fields.

Something had to change. As the pace of globalization accelerated and new companies challenged incumbent leaders, the business units that were now funding R&D refocused the shrinking central laboratories on meeting near-term threats. This only reinforced the narrowing of their vision and mission, with effects that are still being felt today.

The end result was that large companies gradually lost the ability to enter new markets with innovative products they had developed internally. Increasingly, corporations came to view acquisition as the preferred path to new businesses. To a large degree, this shift in strategy paralleled the replacement of visionary leaders with more financially-focused executives.

Assessing potential risk and reward from technology investments is a difficult and intuitive process. A David Sarnoff might relish the challenge and understand the odds when deciding to bless an investment of the magnitude needed to develop color television. But a less intuitive and more financially-focused management is likely to reach a different conclusion.

Of course, RCA was not alone in having a visionary leader. At IBM, for example, Tom Watson's decision in 1964 to launch the enormously risky and costly IBM System/360 project gave the company the tool it needed to dominate computing for years to come. The System/360 went on to become the foundation for the company's success.

In our own time we have seen Bill Gates drive Microsoft from a small start-up with modest funding to a dominant force in many aspects of computing, based on new product launches. His counterpart at Apple, founder Steve Jobs, has returned from exile to revitalize the company's computer business with a risky decision to jettison its legacy operating system and substitute one based on UNIX. Jobs also revolutionized the music business with his iPod player and iTunes music download service.

On the debit side of the ledger, however, there were too many laboratories that failed to produce highly visible and profitable contributions to their parent corporations. The immensely innovative Xerox Palo Alto Research Center (PARC) is often singled out (perhaps unfairly) as an example of this phenomenon.[9] Whatever the cause, it was apparent by the 1970s that corporations were moving away from involvement in long-term industrial research projects.

Federal funds helped to fill the gap by supporting research which opened a number of important new fields and ultimately created whole new markets. Satellite-based services are just one example.

The Department of Defense funded the development of new technologies for military applications, producing many different kinds of semiconductor devices now used in computers, communications, or microwave systems (radar). The Defense Advanced Research Projects Agency (DARPA) and the National Space and Aeronautics Agency (NASA) backed R&D on electronic devices of all types, including integrated circuits and lasers, which eventually found their way into commercial products.

But the era of large-scale corporate innovation was drawing to a close. The Westinghouse Research Laboratory disappeared by the end of the 1970s as the company transformed itself. The biggest event, however, was the 1984 breakup of AT&T and the consequent splitting up of Bell Laboratories, the world's largest central laboratory.

The Bell Labs breakup was a watershed event in the decline of corporate research, even though its immediate cause was a government action against its parent company rather than a shift in business strategy. A summary will show how it fits in with the larger trend.

Under the agreement with the government, AT&T initially removed itself from local telephone service. This meant spinning out the Regional Bell Operating Companies (RBOC) as separate entities. To meet the technology needs of the RBOCs, AT&T moved a section of

[9] M. A. Hiltzik, *Dealers of lightning: Xerox PARC and the dawn of the computer age* (New York: HarperCollins Books, 2000).

the Bell Labs staff into a new research entity called Bellcore, which built and maintained operating support software for the local carriers.

The next split-up of Bell Labs staff came after AT&T spun out Lucent Technologies, Agere Systems, and Avaya Communications. Lucent was formed as a network equipment company, Avaya was an enterprise equipment and networking provider, and Agere became a semiconductor components company. Whole sections of the Labs were assigned to the newly-created businesses, and their individual missions were adapted to their owners' needs. For example, about seventy Bell Labs researchers ended up as Avaya's research department. The days of self-selected research projects were at an end.

RCA Laboratories underwent its own radical transition. When General Electric acquired RCA in 1986, the Labs (later renamed Sarnoff Corporation) became a subsidiary of SRI International, a major not-for-profit research institute. It now had to find all its funding from outside sources.

## Research in the post-Edison era

There are still some large technology companies in the United States, Europe, and Japan that fund central industrial laboratories. The programs of these labs, though strongly guided by the needs of the parent company, are more exploratory in nature than those of the product development organizations attached to business units. Their basic principle is to avoid investing in projects that their parents are unlikely to commercialize. For example, the head of the Sun Microsystems Laboratory defines his organization's mission this way: "To solve hard technical problems brought to us by our customers."[10]

The most prominent of today's corporate labs include those at GE, IBM, Microsoft, Qualcomm, and Corning's Discovery Center, to name a few funded by US corporations, but they are no longer located exclusively in the US. There are now branch laboratories in China and India. Central laboratories exist in Japan and Europe as well. The resources they earmark for exploratory work is generally less than 10 percent of their total R&D budget.

However, these labs are the exception. The Edisonian concept of a central research laboratory still has value, but for the most part

[10] G. H. Hughes, "Sun's R&D spectrum," *Computerworld* (June 5, 2005), 29.

corporations are unwilling to support its activities. The focus of senior management within many large companies faced with pressing investment needs from all sides is likely to be financial rather than technical. They only want innovations that address immediate business needs. This has led them to move research organizations out of the corporate sphere and into the business units, where their proximity to the market will presumably give their activities a sharper commercial focus.

Business units are not congenial places for long-term research. They won't tolerate slow delivery of even the most promising innovations. They are also more interested in extensions to their current product lines than in developing new businesses, and so are more likely to fund safe projects that address current markets than risky innovations, no matter how promising. In times of poor financial performance they will more than likely back off on funding advanced R&D regardless of its relevance to their business.

The 1970s and the breakup of Bell Labs marked the end of an era. For over 50 years central corporate laboratories had been primary sources of revolutionary and evolutionary innovations. Their achievements helped create the digital revolution, just as Edison's inventions built the foundation for the Industrial Age.

But the time of these great centers of commercial innovation seems to be over. Today, when large companies decide to enter a new business, they commonly go outside to buy the necessary technology. Their internal research organizations focus on current business needs.

In the next few chapters we will explore how new technology, new organizational structures, and new funding methods have sprung up to provide viable alternatives to the role of the corporate laboratory in the electronics industry. It will come as no surprise that the basic processes of R&D have not varied much since Edison's time. What has changed is not so much what researchers do or how they do it, but where they do it, the organizations behind them, and the way they realize the economic value of their creations. We will see small entrepreneurial groups under visionary leadership developing the products and services of the future.

At the end of the day, innovation comes down to the creativity of individual human beings with good ideas and a path to the market. That much, at least, will not change.

# 4 | *R&D goes global*

Innovation is an historic and irreversible
change in the way of doing things ... This
covers not only techniques, but also the
introduction of new commodities, new
forms of organization, and the opening of
new markets ... Major innovations entail the
construction of a new plant and equipment,
but not every new plant embodies an
innovation.[1]

We must distinguish, says Schumpeter,
between innovation possibilities and the
practical realization of these possibilities.
Prosperity does not arise merely as a result of
inventions or discoveries. It waits upon the
actual development of innovations which is
the driving power of the period of
prosperity.[2]

Technology, like life, evolves in stages. The genesis of digital electronics
was a handful of simple yet profound innovations, the silicon counterparts of single-celled organisms in primordial seas. Soon successive generations of the technology, each more complex than the last, appeared at
unheard-of speed and spread across the face of the earth.

Unlike the progenitor of life, the creative spirit behind the digital
electronics revolution is easy to identify. It is the genius of innovation,
embodied in the visionary researchers who developed solid-state electronics, ICs, sensors, and software.

---

[1] J. A. Schumpeter, *Readings in business cycle theory* (Philadelphia: The Blackiston
Company, 1944), p. 7.
[2] A. H. Hansen, "What is industrial innovation?" *Business cycles and national
income* (New York: W.W. Norton and Company, 1951), p. 301.

Fifty years ago these innovators were concentrated within the pioneering R&D organizations of a few American corporations. They invented the transistor, fiber optics, the microprocessor, and the semiconductor laser. They created new software structures and languages.

In Schumpeter's sense of the term, these were *revolutionary innovations*, with far-reaching (and largely unforeseen) consequences. By their transformation of electronic technology, they gave rise to entirely new industries.

They were followed by a cascade of *evolutionary innovations* which translated the radical concepts of digital technology into marketable products and services. Researchers developed methods and tools to manufacture the basic components. They designed computers and other products and systems to use them, created networks and infrastructure to extend their functionality, and made the technology affordable to the masses.

The success of all these innovations is self-evident. From calculators to PCs, from CDs to iPods, from cell phones to vast computing and communications networks, digital electronics are now as central to our life as electricity.

Their economic impact has been just as pervasive. In 2004, according to the Consumer Electronics Association, consumer expenditures on digital devices and services in the US alone totaled nearly a quarter of a trillion dollars.[3]

While electronic technology was becoming a crucial component of the world economy, a less visible but equally revolutionary change was under way. Many of the evolutionary innovations that helped spur the growth and acceptance of digital electronics have come from countries other than the US.

Innovation, like electronics, is spreading to the ends of the earth. First its products, and now its production are taking on an international character.

In part this is due to the explosive growth of overseas manufacturing. When factories started to appear in countries with lower costs, local

---

[3] From "A truly digital America," www.ce.org/Press/CEA_Pubs/827.asp (accessed on February 2, 2006).

engineers learned more than manufacturing skills. They developed knowledge of the new technologies and began to produce evolutionary innovations of their own.[4]

Overseas outsourcing and industrial enterprise were not the only factors helping emerging economies develop innovation capabilities. As we shall see, governments have also been actively promoting industrial development, and establishing R&D centers to support this strategy.

This is a logical development. Innovation has always been the most effective way to create new prosperity, new companies, and new high-wage jobs. The electronics industry is just the latest and most spectacular example. Any country that's serious about growing its economy will try to foster local innovation so it, too, can reap the social and economic benefits.

Wen Jiabao, premier of China, made the case for domestic innovation very succinctly. He said it was "impossible to buy core technology; the national strategy is independent innovation."[5]

To achieve this goal, governments are directly or indirectly funding innovation activities in cooperation with industry and, increasingly, with universities.

In the following pages we will look at how government and university support for R&D is being implemented in countries around the world, including the US. We will also appraise the effectiveness of these strategies, and speculate on their ultimate significance for the future of industrial progress.

To support this discussion we will summarize the most significant developments in the globalization of R&D, including:

• the global fostering of technical education;
• the worldwide dissemination of technical talent and its relative quality;
• R&D investments around the world and how they are being allocated among various disciplines;
• government support for industry, especially through national and regional laboratories;
• migration of R&D from developed to developing countries, primarily China and India;
• increases in research outcomes from universities around the world;

---

[4] See the account in C. Perez, *Technological revolutions and financial capital* (Cheltenham, UK: Edward Elgar, 2005).
[5] Quoted in D. Normile, "Is China the next R&D superpower?" *Electronic Business* (July, 2005), 37–41.

- examples of university-industry collaboration in developing new products for the market;
- internationalization of patent filings – an indication of the growing innovative capabilities of developing economies.

Surveying trends in global R&D is a daunting task, involving a blizzard of facts and statistics from dozens of sources. But it will provide a context for discussing some important topics. We will consider how the ultimate commercial value of R&D investments can be assessed, and whether the R&D activity in developing economies is a threat to the leadership of the developed countries in industrial innovation.

## High-tech industries as a national priority

History teaches us that industrial innovation is hardly a new phenomenon. In fact, the development of a high-tech industry follows a basic, replicable pattern, refined over hundreds of years.

All technology industries create products based on the discovery of specific materials, processes, or procedures. They export their products at high profit margins to countries that lack the knowledge, materials, or tools to make them.

A technology industry delivers rising real wages and an improved standard of living for the people in its home country as long as it remains competitive. That's why workers, owners, and government officials alike all find new industries so attractive. But when other countries establish competitive industries, the monopoly profits erode, and the industry is no longer a dynamic creator of prosperity.

For example, the textile industry of medieval Flanders, the fine glass industry of thirteenth-century Venice, and the Chinese porcelain industry prior to the eighteenth century were the high-profit envies of their age. Each dominated its market for a century or more. All three eventually declined as others copied their products.

It may seem odd to talk about centuries-old industries as high-tech. However, Paul A. Krugman has defined a technology industry as "one in which knowledge is a prime source of competitive advantage for firms, and in which firms invest large resources in knowledge creation."[6] Seen this way, every new industry is high-tech for its time.

[6] P. R. Krugman, "Technology and international competition: A historial perspective," *Linking trade and technology policies: An international comparison of the*

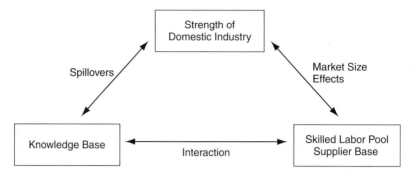

**4.1.** The self-reinforcing advantage of high-tech industries. Reprinted with permission from *Linking trade and technology policies: An international comparison of the policies of industrialized nations.* Copyright © 1992 by the National Academy of Sciences, courtesy of the National Academies Press, Washington, DC (ref. 7).

Krugman's expanding knowledge base is the most important prerequisite for building a successful domestic high-tech industry. The other necessary elements, as summarized in Figure 4.1,[7] include capital investment, a base of suppliers, and a skilled labor pool to make the products.

Knowledge is a competitive advantage, which is why industries keep the processes behind their innovations as secret as they can. Sooner or later, however, by fair means or foul, competitors learn the secrets of each industry's technology and duplicate its products. At this point the industry's monopoly profits melt away.

The modern term for this phenomenon is "technology migration." It used to take centuries, as in the case of the Venetian glass industry. Secrets could be kept for a long time in a world where travel and communication were far more difficult than they are now.[8] Things went faster during the nineteenth century, when Germany became an industrial power.

But in today's wired world, technology migration happens at extraordinary speed. Japan provides the most prominent example in our

  *policies of industrialized nations* (Washington, DC: National Academies Press, 1992), p. 13.
[7] Krugman, *Linking trade and technology*, p. 15.
[8] For an interesting review of early technology migration, see A. C. Crombie, *The history of science from Augustine to Galileo* (New York: Dover Publications, 1995).

time. In less than four decades it rose from the wartime devastation of the early 1950s to become the world's second-largest economy.

Japan started its resurgence as a low-cost manufacturing site for labor-intensive industries, including electronics. Japanese companies then licensed American innovations, manufactured them, and marketed them domestically and abroad. They added evolutionary innovations of their own to make the products more attractive, and the Japanese economic miracle was under way.

Perhaps the most noteworthy factor behind that success is the Japanese focus on innovative manufacturing management and related technologies, which allowed them to cut costs while increasing quality.

## Smaller investment, easier migration

Several of today's fast-growing economies in other parts of Asia also started as low-cost manufacturers in the 1970s, and are striving to emulate Japan's example. South Korea, Taiwan, Singapore, India, and China have made their impact even faster than Japan was able to manage, for reasons that have a lot to do with the current state of digital technology.

For one thing, the price of entry into the electronics industry is lower today than it was fifty years ago. R&D programs in the early days of solid-state electronics had to focus on finding new electronic materials with properties that could be exploited to create new devices.

This approach was prohibitively expensive. It takes sophisticated facilities and on-site collaboration among many specialists to produce commercially viable results from materials-based research. The few large corporate laboratories that dominated early electronic innovation had an advantage: their companies could afford it.

Materials-based research is still expensive. As the materials and processes under investigation grow increasingly exotic, the equipment to carry out the research also becomes more and more costly.

Fortunately, an enormous amount of basic research in materials has already been done. Newcomers to the electronics industry have avenues to innovation that do not require a large capital investment. The most important of these new points of entry is software-based research. All you need to get started are computers and expertise. Both are cheaper than growth chambers for crystal structures.

There are functional advantages to this approach as well. First, bringing expertise together is no longer a problem, because software code doesn't require a lab. The ready availability of high-speed communication links lets collaborators participate in "hands-on" work from remote locations. Project leaders can choose the best people for the job no matter where they're located.

Software-based research also helps level the playing field in favor of companies or countries with limited resources. In software development a small team of outstanding individual contributors will often out-produce a large, cumbersome corporate staff. Access to unlimited resources does not confer an automatic competitive advantage. In this field, quality trumps size.

### Software as equalizer

The rise of software innovation affects all electronics, because software now drives hardware. Integrated circuit design, for example, is a software process. It can be carried out by a team of design experts working at inexpensive computers. Manufacturing the physical ICs, on the other hand, requires costly production plants, of which there are only a few dozen in the world. Thousands of small organizations can design chips, but only a few manufacture them.

This situation has enormous implications for the future of innovation. With software as a central factor in developing new products and services, in a very real sense we have returned to the early years of the Industrial Revolution, when extraordinary creative individuals could seed new industries. It's the Edison model in its purest form.

The ascendancy of software-based research is a boon to developing countries as they try to start industries of their own. They have easier entry to the electronics industry, and their low labor costs give them an advantage. All they need is creative individuals to produce new innovations.

And that's where their advantage stops. The US has maintained a consistent lead in software-based research because of its vast depth of experience, and because it provides a rewarding environment for small teams of truly creative software developers. There are creative individuals in the developed world, too, and they have the benefit of more support, equal or superior expertise, and a lot of hard-won experience.

As we review the strategies that developing countries use to encourage innovation, it is crucial to keep this simple fact in mind: innovative research occurs most often in an unregimented environment, where open inquiry is encouraged and collaboration across functional boundaries is a priority. These have always been the essential cultural elements of the good research laboratories.

If I keep insisting on the importance of individual initiative in productive research, it is because I have seen what happens in regimented organizations. When managers introduce bureaucratic control procedures and arbitrary timelines in an attempt to improve productivity, innovative thinking is stifled and projects fail.

Innovation can succeed only when researchers are allowed to play out their insights to the end. Some of these ideas succeed, others fail, but that is how great results are eventually achieved.

In effect, I believe that the great laboratories of the past were built on a very special culture which cannot be replicated by government fiat. Hence, attempts to build massive research organizations *rapidly* are not a recommended path to building a productive R&D effort.

## Where government support helps

The rapid development of countries such as Germany in the nineteenth century, and Japan in the twentieth, taught economic planners a useful lesson: governments must finance the framework for industrialization. This is one area where government involvement in R&D produces positive results.

The US explored this approach in the mid-1920s. At the time the country had limited experience with ambitious government programs, and the Republican Party was even more suspicious of government "meddling" in the economy than it is now. Yet Secretary of Commerce Herbert Hoover, a Republican, called for government support for basic research. He was concerned that applied science, and hence industrial innovation, would prove unable to sustain itself solely on the basis of business investment.[9]

Many governments have followed suit, anxious to see their electronics industries move upstream in product value. To help the transition

[9] D. O. Belanger, *Enabling American innovation* (West Lafayette, Indiana: Purdue University Press, 1998), p. 17.

from manufacturing resource to technology originator, they have focused on four basic objectives.

1. Increase the population of highly technically trained people.
2. Fund local R&D to reduce the reliance on imported technology.
3. Attract high-tech industries with subsidies and trade benefits.
4. Encourage access to the local market through joint manufacturing ventures with technology-rich foreign companies.

The methods for achieving these ends differ from country to country. This chapter looks at how governments around the world approach the first two objectives, starting with the education of technical talent. We'll leave the last two for our survey of manufacturing in Chapter 6.

## Expanding the global talent base

Since every industry needs a skilled labor pool to draw on, countries seeking to industrialize always try to build an educated workforce. Getting students to take up courses of study in technology has never been a problem for these countries. Many university students willingly choose to pursue a technical education in a growing economy.

One attraction is the ready availability of jobs in technical fields. Technology industries also hold out the promise of a career path based purely on talent, which is available in few other occupations.

In spite of the lure of technical careers, however, the number of students studying science and engineering varies greatly from country to country.

Figure 4.2 ranks member countries of the Organization for Economic Cooperation and Development (OECD) by the percentage of college degrees awarded in science and engineering.[10] The figures range (for 2001) from about 10 to nearly 40 percent. Overall, science-related degrees in OECD countries average 23 percent of the total degrees granted.

South Korea, where nearly 40 percent of all degrees are in science and engineering, is clearly the champion. It is no accident that Korea is one of the countries that has industrialized the fastest over the past thirty years. The US is sixth from last at 16 percent.

---

[10] OECD, *Science, technology and industry outlook* (Paris: OECD Publications, 2004), p. 150.

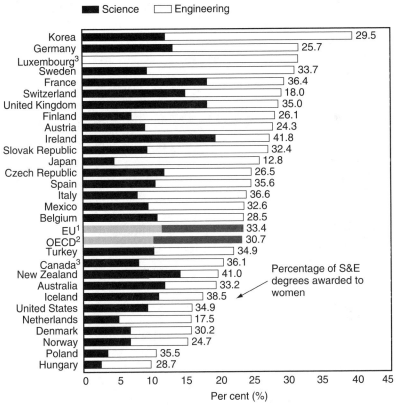

Science ☐ Engineering

4.2. Degrees awarded in science and engineering (S&E) as a percentage of new degrees in 2001 (from OECD, *Science, technology and industry outlook*, ref. 10).

We must be careful about drawing conclusions from these figures, however. The fact that a country produces large numbers of technical graduates does not mean that its economy can absorb them all. Depending on the degree of industrial development in each country, there can be either a surplus or shortage of technical talent.

For example, in 2001 China produced 600,000 technical graduates, while India accounted for 350,000. In the same year the comparable figures for the US and Japan were about 118,000 and 134,000, respectively.[11] China and India are clearly producing surplus engineers, while the US must import talent.[12]

## Quality and quantity: China and India

Nor do these figures tell the whole story regarding the actual level of technical talent available in a given country. India and China are excellent cases in point.

India's top technical schools include the seven Indian Institutes of Technology (IIT), which accept less than 1 percent of applicants. Together these institutions produce a few thousand of the world's best-trained engineering and science graduates every year.

However, this accounts for only about 1 percent of the country's total. Furthermore, many of the elite graduates emigrate to other countries. (We examine this phenomenon in more detail later in this chapter.)

At the other end of the quality scale are some of India's new private schools, opened to fill the booming demand for software engineers. In 2005, nearly 100 such institutions were reported closed by the Indian Supreme Court because of unacceptable quality standards.[13]

The wide disparity between IIT and other Indian colleges is a symptom of high demand overwhelming limited supply. While training qualified software engineers requires little capital investment, it does presuppose qualified teachers. These are a scarce commodity, because in emerging economies, industrial jobs often pay much better than teaching, particularly at less prestigious institutions. The resulting shortage of teaching talent is a huge barrier to the expansion of technical education.

China's educational system presents, if anything, an even more contradictory picture. On the negative side, a 2005 McKinsey report finds

---

[11] OECD, *Science, technology and industry*, p. 218.

[12] Many experts worry that we are overreliant on foreign sources of talent. See G. Colvin, "America isn't ready [Here's what to do about it] in the relentless, global, tech-driven, cost-cutting struggle for business," *Fortune* (July 25, 2005), 70–77.

[13] "The brains business: A survey of higher education," *The Economist* (September 10, 2005), 15.

that only a fraction of the country's graduates are believed to be sufficiently trained to work in the high-tech export sector. The reasons for this situation are narrow training and inadequate English language skills.[14]

Yet there is no question that China is rapidly expanding its pool of high-quality talent. The ranks of elite Chinese universities aspiring to world-class status have grown significantly. They include Tsinghua University (Beijing), Zhjiang University (Hangzhou), Shanghai Jiaotong University, Hua Zhong University of Science and Technology (Wuhan), University of Science and Technology of China (Hefai, Anhui Province), and Hunan University of Technology (Guangzhou).

These institutions combine high-quality teaching with research activities along the lines of the best American universities. As another sign of progress, the number of doctoral degrees awarded by Chinese universities has increased rapidly, from about 2,500 annually in 1990 to about 18,000 in 2003.

In spite of this impressive progress, Indian university graduates still enjoy one major advantage over their Chinese counterparts in the world market for technology – their training is in English. The success that India has had in building a software service industry to meet the needs of US customers can be attributed largely to the language skills of the engineers.

The net result of the growing pool of engineers around the world is a huge disparity in their cost, including corporate salaries and overheads. Table 4.1 (based on my personal communications with companies with facilities in various geographies) shows estimates in 2006 of the fully burdened cost of employing engineers in different countries.

## Brain drain, brain gain

From the point of view of emerging economies, education is a domestic necessity and a foreign threat.

Countries must build a solid educational system if they are to keep their brightest students at home to help build high-tech industries. If they do not, top-flight universities in other countries will attract their most ambitious and talented prospects. The danger is that if these

[14] G. Dyer and K. Merchant, "Graduates may fail Chinese economy," *Financial Times* (October 7, 2005), 8.

*Table 4.1 Annual engineering costs*[a]

| Country | Entry level ($) | 5-year experience ($) |
| --- | --- | --- |
| China | 5,000 | 9,000 |
| Ukraine | 6,000 | 8,000 |
| India | 20,000 | 25,000 |
| Malaysia | 21,000 | 27,000 |
| Mexico | 21,000 | 27,000 |
| Korea | 22,000 | 32,000 |
| Finland | 40,000 | 60,000 |
| Israel | 60,000 | 80,000 |
| Canada | 60,000 | 80,000 |
| United States | 100,000 | 150,000 |

Note:

[a] Fully burdened costs

technical elite find greater rewards in the host country's industries, they may well decide not to return.

The brain drain, is, of course, a reality. Countries with outstanding universities benefit greatly from the enrollment of foreign students, particularly for higher education. These total 1.9 million in the countries listed in Figure 4.3.

As shown, 28 percent of the student expatriates were in the US in 2002. By one reliable estimate, they make up an amazing 60 percent of graduate students in the sciences and engineering. The UK accounted for 12 percent of the expatriate total, followed by Germany with 11 percent, France with 10 percent, and Australia with 9 percent.[15]

The brain drain from the students' countries of origin creates a brain gain for their host countries. In its survey of global higher education,[16] *The Economist* provided an estimate of the number of highly-skilled immigrants who are pursuing careers in developed countries. The US, with 8.2 million, has the largest number of any country by far. You would have to combine the figures from the next four countries on the list (Canada, Australia, Britain, and Germany) to get a comparable total.

[15] *The Economist*, "The brains business," 16.
[16] *Ibid.*, 18.

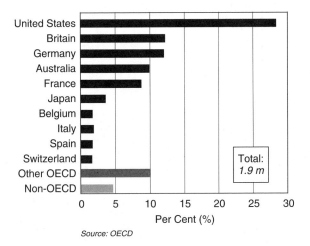

**4.3.** Percentage of foreign students in tertiary education by country of study in 2002. Copyright © 2005 The Economist Newspaper Ltd. All rights reserved. Reprinted with permission. Further reproduction prohibited; www.economist.com (ref. 15).

A disparity this large calls for some explanation. We noted above that the US is not producing enough engineers to fill its needs, especially in disciplines where there is high demand for skilled practitioners. The figures from *The Economist* reinforce what other studies have shown: to a significant extent the US is filling the gap with talented foreign students who remain in the country after their graduation. Among foreign PhDs who completed their studies in the US between 1998 and 2001, about 70 percent are believed to have remained in the country with permanent jobs.[17]

The US need for foreign talent is especially great in information technology. A study quoted by David A. Patterson[18] (see Figure 4.4) shows that the interest in computer science on the part of American college students is cyclical, and it is currently on the downswing. In 1982 some 5 percent of incoming freshman listed computer

[17] *Ibid.*, 21.
[18] D. A. Patterson, "President's letter – Restoring the popularity of computer science," *Communications of the ACM* 48 (September 2005), 25; derived from data from the CIRP Freshman Survey on entering students' interest in computer careers from the Higher Education Research Institute (HERI) at UCLA.

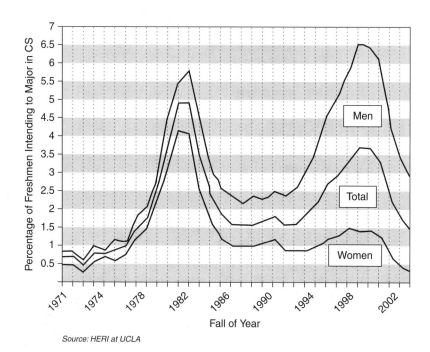

Source: HERI at UCLA

**4.4.** Computer science (CS) listed as probable major among incoming freshmen since 1971 in the US. Reprinted from "President's letter," *Communications of the ACM*, 48:9, September 2005. Figure reprinted with permission from the Higher Education Research Institute (HERI) (ref. 18).

science as a potential major. This number was down to 1.5 percent in 2004.

One explanation, offered by Patterson, is that students are avoiding the IT profession because news about jobs being sent to India has convinced them there is no future in the field. This is probably an over-reaction to bad news.[19] As shown in Figure 4.5, US Department of Labor statistics indicate that domestic IT employment has grown by 17 percent since 1999. It rose to over three million workers in 2004 despite the shift of IT engineering jobs to India during this period.

[19] Patterson, "President's letter," 26.

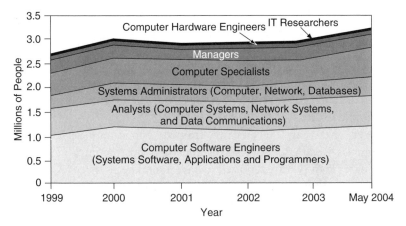

Source: National Bureau of Labor Statistics

**4.5.** Annual information technology employment in the US from 1999 based on US Bureau of Labor Statistics. Reprinted from "President's letter," *Communications of the ACM*, 48:9, September 2005 (ref. 19).

## Reciprocal benefits

If Americans are unwilling to bet their futures on careers in information technology, foreign professionals are more than willing to take their place. About 400,000 science graduates from the EU were residing in the US in 2005, and only about 14 percent are prepared to return home, according to a survey quoted in the *Wall Street Journal*.[20]

Since their presence helps sustain the US industrial base, the government has established a procedure to accommodate them. Special H-1B visas allow an annual immigration of engineers to fill jobs for which local talent is not available. There were about 65,000 in 2006. There has been no problem filling these quotas.

Some immigrants are attracted by the technical environment. The *Wall Street Journal* article cited above quotes a Russian software engineer as saying: "This is the original country for computer science. This is the ultimate place."

Another draw is the belief that a successful US residency is a validation of the immigrant's professional standing. An Indian consultant

[20] J. Kotkin, "Our immigrants, their immigrants," *Wall Street Journal* (November 8, 2005), A16.

remarked, "There is no dearth of jobs for the qualified in India, but a US job is quite another thing. It's seen as a career landmark, like a prestigious MBA, that guarantees strong career growth."[21]

Such immigrants contribute enormously to their adopted countries through their entrepreneurship, their technical achievements, or their work as university faculty. But their biggest impact occurs in the economic realm.

Over the years, technically trained immigrants have created enormous wealth for themselves and their host country as well. IT graduates in the US are reported to have a total net worth of $30 billion dollars. It is safe to assume that the value of enterprises which they helped create is many times that number.[22]

## R&D investments around the world: An overview

There's little benefit in creating a pool of highly trained people if they have nothing to do. In addition to educating technical talent, governments around the world are also engaged in ambitious efforts to fund and encourage R&D on a national level.

The more developed economies see ongoing research as vital to supporting and advancing their industrial base. For countries with emerging economies, local research not only helps create new products and services, it reduces dependence on imported technology.

Table 4.2 shows the total national R&D expenditures (normalized to 1995 dollars) of various countries in 1981 and 2002, according to the OECD.[23] Total global R&D spending reached about $657 billion in 2002.

Because this figure included corporate and government-funded expenditures, which support defense, academic, and industrial programs, inter-country comparisons are hard to calibrate in terms of economic value. The notoriously elastic definition of R&D makes the problem worse. Still, these data show a significant increase in national expenditures to support technology development over the twenty-one-year period.

---

[21] P. Thibodeau, "IT groups push congress to raise H-1B visa limits," *Computerworld* (October 3, 2005), 12.

[22] *The Economist*, "The brains business," 18.

[23] OECD, *Science, technology and industry*, pp. 190–191.

In 2002 the US spent $245 billion, followed by Japan with $94 billion, then China with $65 billion. The number for China includes expenditures by foreign corporations. We can expect this to grow as more foreign companies move engineering facilities into China.

Absolute dollar numbers don't allow us to compare manpower figures, as salaries are so different from one country to another. R&D expenditures as a fraction of the gross national product (GDP) are perhaps more meaningful. As Table 4.2 also shows, China's outlays increased from 0.74 percent of GDP in 1981 to 1.23 percent in 2002.

By comparison, R&D in the UK declined from 2.38 percent of GDP in 1981 to 1.88 percent in 2002. Japan, which trailed the US in 1981, moved to the top of the rankings among major economies by spending at least 3.12 percent of GDP, while the US reached a level of 2.67 percent.

The growth in China's spending level has not escaped the attention of European Union officials, who are charged with maintaining employment. To quote Janez Potocnik, the EU Commissioner for Research: "The Chinese trend is extremely clear. If the trend continues, [China] will catch up with us in 2009 or 2010. It is obvious that [European companies] are transferring some of their investments there."[24] However, the Commission appears to have little ability to get the EU to adopt a more aggressive inward investment strategy.

So far we have discussed *total* R&D spending in each country, including government and private sectors. Let's now consider *government* R&D spending on its own, to get a better picture of each country's innovation priorities.

Government budget appropriations for various broadly defined objectives (on a percentage basis) for the years 1991 and either 2001 or 2003 are shown in Table 4.3.[25] Note that in the latter years the US, UK, and Russian Federation devoted the largest shares of their budgets to defense, whereas Korea and Japan spent the highest percentages on economic development.

The US government spent about $100 billion on R&D in 2003,[26] but the bulk of these funds went to defense initiatives (54%) and

---

[24] R. Minder, "Chinese poised to outstrip Europe's spending on R&D," *Financial Times* (October 10, 2005), 4.

[25] OECD, *Science, technology and industry*, p. 213.

[26] Statistics from www.nsf.gov/statistics/infbrief/nsf03321/ (accessed on February 2, 2006); also www.aaas.org/spp/rd/fy06.htm (accessed on February 2, 2006).

Table 4.2 National R&D expenditures for 1981 and 2002

| Country | 1981 | | 2002 | |
|---|---|---|---|---|
| | 1981 gross R&D expenditures (M) constant US dollars (1995 value) | 1981 – as a percentage of GDP | 2002 gross R&D expenditures (M) constant US dollars (1995 value) | 2002 – as a percentage of GDP |
| United States | 114,530[l] | 2.34[l] | 245,430[l,o] | 2.67[l,o] |
| Japan | 38,752[i,m] | 2.12 | 94,172 | 3.12 |
| China | 13,824[c,n] | 0.74[c,n] | 65,485 | 1.23 |
| Germany | 27,895 | 2.43 | 48,934[i] | 2.52[i] |
| France | 17,870[b] | 1.93[b] | 31,923[o] | 2.2[o] |
| United Kingdom | 19,201[b] | 2.38[b] | 26,207 | 1.88 |
| Korea | 7,563[c,k] | 1.92[c,k] | 20,858[k] | 2.91[k] |
| Canada | 5,843 | 1.24 | 16,072[p] | 1.91[p] |
| Italy | 7,914[q] | 0.88[q] | 14,830[e] | 1.11[e] |
| Russian Federation | 23,032[c] | 1.43[c] | 13,651 | 1.24 |
| Sweden | 3,234[h,n] | 2.22[h,n] | 9,503[e,n] | 4.27[e,n] |
| Spain | 1,754 | 0.41 | 8,090 | 1.03 |
| Netherlands | 4,304 | 1.79 | 7,670[e] | 1.89[e] |
| Israel | 1,937[c,j] | 2.5[c,j] | 5,516[j,o] | 4.72[j,o] |
| Switzerland | 3,233[i] | 2.12[i] | 5,255[d] | 2.57[d] |
| Finland | 904[b] | 1.18[b] | 4,374 | 3.46 |
| Austria | 1,457 | 1.13 | 4,098[i,o] | 1.93[i,o] |
| Rest of World (ROW) | 17,656[a,b,f,i,j,p] | – | 35,164[g,i,j,o] | – |
| TOTAL WHOLE WORLD | 310,903[i] | – | 657,232[i,o] | – |

**Table 4.2 (*cont.*)**

*Notes:*

Year availability

[a] Some ROW countries 1982 instead of 1981

[b] Some ROW countries 1983 instead of 1981

[c] 1991 instead of 1981

[d] 2000 instead of 2002

[e] 2001 instead of 2002

[f] Some ROW countries 1991 and 1995 instead of 1981

[g] Some ROW countries 2000 and 2001 instead of 2002

Standard statistical notes used for science and technology indicators

[b] Break in series with previous year

[i] Estimate

[j] Defense excluded (all or mostly)

[k] Excluding R&D in the social sciences and humanities

[l] Excludes most or all capital expenditure

[m] Overestimated or based on overestimated data

[n] Underestimated or based on underestimated data

[o] Provisional

[p] Does not correspond exactly to the OECD recommendations

[q] Including extramural R&D expenditures

*Source:* OECD, MSTI database, May 2004, reproduced in OECD, *Science, technology and industry outlook* 2004, © OECD 2004.

Table 4.3 *Government budget appropriations and outlays for R&D by socioeconomic objectives, 1991 and 2003 as a percentage of total R&D budget (percentages do not always add up to 100% because of use of dissimilar years in some cases)*

| Country | Defense 1991 | Defense 2003 | Civil Economic development 1991 | Economic development 2003 | Health 1991 | Health 2003 | Space 1991 | Space 2003 | Non-oriented programs 1991 | Non-oriented programs 2003 | General university funds 1991 | General university funds 2003 |
|---|---|---|---|---|---|---|---|---|---|---|---|---|
| Australia | $10.3^b$ | $5.7^{b,m}$ | $25.8^b$ | $34^{b,m}$ | $14.5^b$ | $19.9^{b,m}$ | 0 | $0^{b,m}$ | $15.0^b$ | $3.7^{b,m}$ | $34.4^b$ | $36.4^{b,m}$ |
| Austria | $0^b$ | $0^{b,m}$ | $14.6^b$ | $12.7^{b,m}$ | $8.6^b$ | $8.5^{b,m}$ | $0.4^b$ | $0.1^{b,m}$ | $12.4^b$ | $13.1^{b,m}$ | $64.0^b$ | $65.5^{b,m}$ |
| Belgium | 0.2 | $0.4^m$ | 25.6 | $36.9^m$ | 10.1 | $9.6^m$ | 12.4 | $8.9^m$ | 22.7 | $22.9^m$ | 23.9 | $18.2^m$ |
| Canada | $5.1^b$ | $4.3^{a,b}$ | $33.8^b$ | $32.0^{a,b}$ | $13.8^b$ | $23.5^{a,b}$ | $7.2^b$ | $6.2^{a,b}$ | $12.5^b$ | $7.2^{a,b}$ | $27.6^{f,b}$ | $25.7^{a,f,b}$ |
| Denmark | 0.6 | 1.1 | 26.3 | 16.5 | 14.1 | 16.7 | 2.7 | 2.2 | 23.3 | 20.6 | 33.0 | 42.1 |
| Finland | $1.4^e$ | $2.9^m$ | $40.4^e$ | $39.1^m$ | $16.3^e$ | $15.2^m$ | $3.1^e$ | $1.9^m$ | $10.5^e$ | $13.7^m$ | $28.3^e$ | $27.2^m$ |
| France | 36.1 | $24.3^{b,m}$ | 21 | $12.3^{b,m}$ | 6.3 | $10.2^{b,m}$ | 8.6 | $8.9^{b,m}$ | 15.3 | $19.7^{b,m}$ | 12.4 | $23.0^{b,m}$ |
| Germany | $11.0^e$ | $6.7^m$ | $22.7^e$ | $19.1^{m,i}$ | $11.6^e$ | $13.7^{m,i}$ | $5.4^e$ | $4.9^{m,i}$ | $15.2^e$ | $16.6^{m,i}$ | $33.2^e$ | $39.3^{m,i}$ |
| Italy | 7.9 | $4.0^{a,m}$ | 21.8 | $16.1^{a,m}$ | 18.2 | $15.5^{a,m}$ | 7.0 | $7.3^{a,m}$ | 10.6 | $13.3^{a,m}$ | 31.3 | $43.7^{a,m}$ |
| Japan | $5.7^{g,b,k}$ | 4.5 | $31.6^{g,b}$ | $31.9^{b,m}$ | $5.1^{g,b}$ | $7.3^{b,m}$ | $6.8^{g,b}$ | 6.7 | $8.0^{g,b}$ | $15.3^{b,m}$ | $42.5^{g,b}$ | $34.4^{b,m}$ |
| Korea | 0 | 14.2 | 0 | 44.7 | 0 | 16.7 | 0 | 2.8 | 0 | 21.6 | 0 | 0 |
| Netherlands | 3.0 | $1.9^a$ | 28.1 | $25.3^a$ | 8.7 | $8.7^a$ | 2.6 | $2.6^a$ | 10.6 | $10.7^a$ | 43.0 | $46.3^a$ |
| Russian Federation | 0 | $43.5^{a,f}$ | 0 | $24.4^a$ | 0 | $7.0^a$ | 0 | $10.1^a$ | 0 | $14.0^a$ | 0 | $0^a$ |
| Spain | 16.8 | $37.3^{a,f}$ | 27.5 | $22.7^{a,f}$ | 15.1 | $9.7^{a,f}$ | 7.0 | $2.4^{a,f}$ | 10.8 | $2.1^{a,f}$ | 20.0 | $25.8^{a,f}$ |
| Sweden | 27.3 | 22.2 | 17.8 | 13.6 | 8.3 | 8.9 | 1.7 | 0.6 | 14.6 | 16.7 | 30.4 | 38.0 |

*Table 4.3 (cont.)*

| | | | | | | | | | | | | |
|---|---|---|---|---|---|---|---|---|---|---|---|---|
| Switzerland | $4.6^{c,b}$ | $0.7^{d,b}$ | $3.7^{c,h,k}$ | $4.6^{d,b,k}$ | $3.5^{c,h,k}$ | $2.4^{d,h,k}$ | 0 | 0 | $0^{c,h}$ | $0^{d,b}$ | $59.3^{c,h,l}$ | $61.1^{d,b,l}$ |
| United Kingdom | 43.9 | $34.1^{b}$ | 16.2 | $9.8^{b}$ | 12.5 | $20.1^{b}$ | 2.7 | $1.9^{b}$ | 5.1 | $13.3^{b}$ | 18.9 | $20.2^{b}$ |
| United States | $59.7^{b,i,j}$ | $53.7^{f,b,i}$ | $8.9^{b,i,j}$ | $5.6^{f,b,i}$ | $17.5^{h,i,j}$ | $26.3^{f,b,i}$ | $9.9^{b,i,j}$ | $8.4^{f,b,i}$ | $4.0^{b,i,j}$ | $6.0^{f,b,i}$ | 0 | 0 |

*Notes:*

Year availability

[a] 2001 instead of 2003

[b] 2002 instead of 2003

[c] 1992 instead of 1991

[d] 2000 instead of 2003

Standard statistical notes used for science and technology indicators

[e] Break in series with previous year

[f] Estimate

[g] Excluding R&D in the social sciences and humanities

[h] Federal or central government only

[i] Excludes data for the R&D content of general payment to the high-education sector for combined education and research

[j] Excludes most or all capital expenditure

[k] Underestimated or based on underestimated data

[l] Includes other classes

[m] Provisional

*Source:* OECD, MSTI database, May 2004, reproduced in OECD, *Science, technology and industry outlook 2004*, © OECD 2004.

health-related research (26%). Only the UK spends comparable percentages on these two activities.

This means that the US government spends only about $30 billion a year on programs unrelated to defense or health technology programs. In constant dollars this amount has not changed much over the past decade.

The $3.2 billion budget of the National Science Foundation, which supports basic physical and computer science research in universities, is part of the non-defense category. It's usually assumed that the spillover from defense programs into commercial applications is modest, but there have been some noteworthy contributions, such as supercomputers, the Internet, and satellite technology.

One agency stands out in this regard: the Defense Advanced Research Projects Agency (DARPA), with about $3 billion in funding. DARPA underwrites the development of multipurpose technologies, and it has a long history of supporting projects that have had a major impact on the industrial sector. DARPA and the National Science Foundation (NSF) have been responsible for funding some farsighted technology developments, including the early innovations that enabled the Internet.

### National research initiatives
Most countries use both indirect and direct research subsidies to encourage their high-tech industries. The primary objective is to develop innovations that will generate thriving local industries.

Indirect subsidies include tax rebates to encourage business spending on R&D. Direct government subsidies include funding for research centers, training facilities, centers promoting the inflow of technology through international relationships, and venture funding of new companies.

A few examples of national research initiatives will illustrate the variety of approaches used by different countries.

### United States: Albany Nanotech Facility
The Albany Nanotech Facility is a regional (rather than national) initiative in support of a highly specialized industry. The facility is funded in part by the State of New York to develop semiconductor manufacturing processes. It is expected to cost about $600 million over seven years, and will serve the advanced technology needs of US-based semiconductor manufacturers such as IBM, Micron Technologies, and

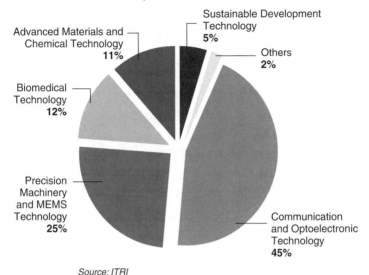

**Revenue: US$506 Million (2003)**

Government Research Project 54%, Industrial Contract Service 46%

*Source: ITRI*

**4.6.** Division of programs at Taiwan's Industrial Technology Research Institute (ITRI) in 2003. Reprinted with permission from ITRI (ref. 28).

Texas Instruments, as well as those of Applied Materials and other suppliers of production equipment for this industry.[27]

### Taiwan: Industrial Technology Research Institute (ITRI)

Established in 1973, ITRI is a government-managed laboratory which serves its country's industries by centralizing certain R&D functions. Its long, successful history merits special attention.[28]

ITRI provides over 600 local companies in Taiwan with many of the services that the big corporate laboratories used to furnish to their parent companies. Offerings range from training programs to the management of technology transfers from foreign sources.

The distribution of ITRI's programs, shown in Figure 4.6, covers both established and developing fields of industrial activity in Taiwan.

---

[27] D. Lammers, "Albany lands lead on litho R&D stage," *EE Times* (July 25, 2005), 1.

[28] Presentation by Dr. P. Lin, Vice President of ITRI (August 26, 2004).

**Total Manpower as of December 31, 2003: 6,193**

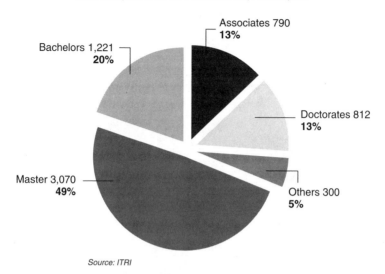

*Source: ITRI*

**4.7.** Educational background of the staff of Taiwan's Industrial Technology Research Institute (ITRI) in 2003. Reprinted with permission from ITRI (ref. 28).

For example, 47 percent of ITRI's revenues from optoelectronic and communications technology go to support research on displays and light emitting diodes, areas where Taiwan has built a world-leading manufacturing position.

Its precision machinery and micro-electro-mechanical systems (MEMS) technology activities, on the other hand, address an emerging industrial area of great importance: new sensor technology that marries mechanical and electronic concepts to create new products.

In 2003 ITRI had a staff of about 6,200 people (including 813 PhDs), and a budget of about $500 million as shown in Figure 4.7. The government provided 54 percent of its funding. The remaining 46 percent came from industrial contracts, under which the organization produces intellectual property for the exclusive use of the funders.

The institute also functions as an incubator for new businesses, and as a training facility for management and technical personnel for local industries.

One of ITRI's most notable achievements was the incubation of Taiwan's semiconductor industry. Starting with a license for CMOS process technology from RCA in 1976, ITRI helped establish the United Microelectronics Corporation (UMC) as well as the Taiwan Semiconductor Manufacturing Company (TSMC).

TSMC became the world's largest and most profitable silicon foundry. ITRI support was instrumental in establishing other companies in various electronics sectors.

To foster an innovative environment for its research staff, the institute frequently hosts foreign researchers and maintains collaborative relationships with research institutions outside of Taiwan. ITRI reports that its staff has filed over 6,000 patents.

It should be noted that university-based research, to be discussed later in this chapter, is funded separately by the National Science Council of Taiwan.

## Hong Kong: Applied Science and Technology Research Institute Ltd. (ASTRI)

Emulating the success of ITRI, Hong Kong's ASTRI aims to promote the development of high-tech industries in Hong Kong and South China. Established in 2001, ASTRI plans to have a staff of 800 in place by 2009, focused on advanced materials research and electronic devices and systems.[29]

The institute's collaboration with overseas researchers fosters an open dialogue on new ideas. It also plans to incubate new businesses on the basis of innovations developed by its staff.

## Singapore: Institute for Infocomm Research (IIR)

Funded by Singapore's Agency for Science, Technology and Research, IIR aims "to benefit humanity and create prosperity for Singapore." With a staff of 350 researchers (in 2004), the institute focuses on combining research into electronic systems and software from local universities with practical development activities to facilitate transfer to industry. The primary objective of this activity is the creation of new industries in Singapore on the basis of local innovations.[30] It will take

---

[29] Presentation by Dr. R. J. Yang, CEO of ASTRI (September 2005).
[30] Communications from Dr. K. Lai, Director, New York Office, Singapore Economic Development Board (November 2004).

years to determine whether this kind of institute can produce the desired results.

### Israel: Support for innovative young companies

Observant readers of Table 4.2 will have noticed that Israel's gross R&D expenditures for 2002 were $5.5 billion. This is an unusually high figure for such a small country, representing 4.72 percent of GDP. It is the result of a forty-year-old government policy to develop a high-tech industrial base.

Six large world-class universities provide Israel with an abundant source of highly-trained technical graduates. Special funding through the office of the Chief Scientist supports the establishment of innovative new companies, while funding for their expansion is available from a well-developed venture capital industry. In addition, there is a deliberate policy of moving technologies developed for the defense industries into the industrial marketplace.

As a result of these strategies, Israel's importance as a source of innovation is out of proportion to its size. Of the major foreign R&D facilities it has attracted, Intel's is the largest, but many corporations with headquarters elsewhere in the world also maintain R&D facilities in Israel.

We will discuss the role of venture capital and the commercial success of Israel's high-tech industry in Chapter 5.

### France: Agency for Industrial Innovation (AAI)

The French government has created a number of programs to support local industries. The OECD[31] reports that France devoted a total of €103 million in 2002 to fund industrial research, up sharply from €41 million in 1998.

The newest initiative substantially expands that effort. The Agency for Industrial Innovation (AAI) was created in 2005, and is charged with financing regional innovation clusters.[32] This agency, with an initial budget of €1 billion, will co-finance industrial innovation programs involving local industry, government laboratories, and universities. The goal is to promote industrial growth and employment in

---

[31] OECD, *Science, technology and industry*, p. 92.
[32] L. Chavane, "One billion euros for the Agency for Industrial Innovation," *Le Figaro* (August 30, 2005), 5.

various parts of France. Out of the 105 fields of activity proposed for funding by the regions, the agency selected sixty-seven, a clear indication of the scope of its ambitions.

### Belgium: Interuniversity Microelectronics Center (IMEC)

Although Belgium does not have a significant semiconductor manufacturing industry, it does have a leading microelectronics research laboratory. IMEC,[33] founded in 1984, is an independent institution affiliated with local universities and supported by government grants and industrial research contracts. The range of its activities, and its unique mission, require some explanation.

IMEC's facilities are staffed and equipped to develop commercial grade manufacturing technologies. Its permanent staff of about 900 scientists, engineers, and technicians, plus an additional 400 temporary guest researchers, can prove out new processes on a fully-equipped prototype production line.

In sharp contrast to the programs we have been considering so far, IMEC's objectives include the globalization of sophisticated technology developed at the center. Guest researchers are expected to transfer the technologies developed at the center to their employers when they return home. For example, IMEC-trained staff have facilitated the start-up of new chip production facilities in China.

IMEC's 2004 operating budget of €159 million was partly funded by the Belgian government, which contributed 24 per cent of the total. Contracts from sponsors that gain access to its technology provide the remainder. These sponsors include corporations in Europe, the US, and Asia.

In return for their funding, sponsors also benefit from an activity that is just as important as personnel training. The center performs the kind of materials science research that was one of the trademark contributions of the big central laboratories at IBM, AT&T, and RCA. As we have already noted, materials research is a costly undertaking. It is obviously more economical to share its cost among several funding sources than to pursue it on an individual basis.

---

[33] From IMEC history, www.imec.be/wwwinter/about/en/IMEChistory.shtml (accessed on April 30, 2006).

IMEC's productivity, and its standing in the technical community is indicated by the number of papers its researchers publish on their results – 1,400 in 2004 alone. It is a highly respected institution.

**Commercial contract research: Sarnoff Corporation**
There are a number of independent organizations in the US that conduct research on behalf of government agencies and industrial clients. They include Battelle, SRI International, Midwest Research Institute, Southwest Research Institute, and Sarnoff Corporation.

Sarnoff, a subsidiary of SRI, is the successor to RCA Labs, where I worked as a researcher and manager. I have served on its board for a number of years, and am chairman at the time of writing. The company's unusual charter suggests an alternative approach to fostering innovation and creating new high-tech companies.

Unlike other independent research organizations, including its parent SRI, Sarnoff is a for-profit enterprise. It is also set apart by its client base and commercial focus. While government research accounts for up to 60 percent of contracts, Sarnoff has historically maintained a good balance between government and commercial clients. It has shown uncommon skill in developing marketable technologies and products, a legacy of its decades of experience in creating technologies for RCA to commercialize.

During the nearly twenty years since its separation from RCA, Sarnoff has developed some of the most successful commercial products and technologies in history on behalf of its clients. It was instrumental in creating direct digital satellite TV, the US standard for digital HDTV, radio frequency ID chips, computer vision technology, advanced LCD displays, *in silico* drug discovery algorithms and processors, disposable hearing aids, MEMS sensors, and self-configuring communication networks.

Sarnoff also participated in the founding of more than twenty new companies to commercialize its innovations. In recent years it has established subsidiaries in Asia and Europe to build a global innovation network for its clients. It is still too early to assess the impact of these initiatives, but the company offers an intriguing alternative to government, university, and industrial research.

## The migration of engineering: India, China, and beyond

Countries with emerging economies are not relying solely on government funding to set up R&D centers. As developing countries have built up their innovation infrastructures, they have been able to attract an impressive array of leading international companies to invest in local facilities and engage local technical talent.

A major driver for this migration of engineering work has been the much lower costs in developing countries as compared to those in Western Europe, Japan, and the US.

A few years ago these investments might have been confined to low-cost, labor-intensive manufacturing plants in China, or business service centers in India. More recently, corporations have been building software development centers in India, and product development and engineering facilities in both China and India. There is similar activity in other parts of Asia, though on a smaller scale.

This phenomenon is no longer confined to European and US companies. Engineering is migrating from Japan, South Korea, and Taiwan as well. Here are just a few examples of the trend.

- Intel, which already has 2,000 engineers in India, expects to have about 1,000 in China by the end of 2006.
- Microsoft has about 200 people in its central research laboratory in China. This organization is charged with exploratory research and is not directly affiliated with specific product groups.
- Cypress Semiconductor, a billion-dollar company, will expand its engineering organization in India from 200 engineers to about 700 in 2007. Cypress says that the chip design effort conducted there has produced forty US patents in ten years.[34] Cypress also owns an engineering center in China.
- MediaTek (Taiwan), a leading consumer electronics chip company, has set up an R&D center in Singapore which will have about 300 chip design engineers.
- Motorola had 2,500 software engineers in India in 2005 and plans to add another 1,000.
- General Electric's largest R&D center is in India, with 2,300 engineers involved in product development for areas including power trains and medical instruments.[35]

---

[34] "Cypress plans $10 million Indian R&D center," *Electronic News* (September 19, 2005).

[35] S. J. Binggaman, "Maintaining America's competitive edge," *APS News* (June 2005), 8.

- IBM has thousands of engineers in China and India. The company also works with new businesses around the world in product areas of interest to IBM. In this way, and through a network of venture capital funding sources, it can keep track of new companies with products that it needs and acquire them if they fit the corporate strategy.[36]
- Alcatel of France manufactures telecommunications equipment in China through Alcatel Shanghai Bell (ASB), which is a joint venture with the Chinese government. ASB employs 2,000 engineers. It enjoys the status of a domestic business entitled to government subsidies and such financial benefits as partial support for R&D and preferred access to bank capital. ASB develops many of the products sold by Alcatel worldwide, and manufactures them in China under very favorable cost conditions.[37]
- Siemens has thousands of engineers in China and India involved in both software and equipment development.
- Infosys Technologies, an Indian information technology services company, in 2005 reported plans to hire 6,000 software engineers in China, which would allow it to participate in the IT service business in China as well as do contract work for overseas businesses.[38]

Few issues are as emotionally charged today as the "outsourcing" of jobs. The temperature of the debates over this practice will no doubt rise as engineering is shifted overseas to join the business process functions already being relocated to India.

The availability of trained engineers in emerging economies does allow more value-added engineering to be done at a lower cost than previously possible. This is good news in an age of reduced R&D budgets. In the long run, China and India will also benefit because their engineers are learning state of the art technologies, thus expanding native innovative capabilities.

But this does not mean these countries have caught up with the advanced economies. Developed countries still produce a greater

[36] A. Clark, "Innovation sourcing: Leveraging the global ecosystem," *Venture Capital Review* (Summer 2005), 7.
[37] Presentation to industry analysts by Alcatel spokesperson, Olivia Qiu (2005).
[38] K. Merchant, "Infosys plans to hire 6,000 staff in China," *Financial Times* (October 1–2, 2005), 8.

number of contributors to the more innovative areas of product research and development. These countries have resources that their newly industrializing partners cannot match: the generally higher quality of their universities, the influx of talented foreigners, and the maturity of their research institutions.

## The growing importance of university research

The university has often been portrayed as an ivory tower, insulated from the hubbub of the commercial world, where theoretical studies increase the store of general human knowledge. Tangible innovations that emerge from this bastion of intellectual inquiry are byproducts of its central mission.

This view of university research is no longer tenable. Universities in general, and technical institutes in particular, are increasingly involved in the development of innovative technology, which they license to outside companies or commercialize through venture capital-funded start-ups.

Exactly how critical universities are in building high-tech industrial prowess is difficult to pin down. The facts and figures on funding and productivity are uncertain. But there is enough to suggest the importance of their role.

### Funding the research

University operations in most countries are fully supported by their national governments.[39] Table 4.3 (shown earlier) shows that in many countries expenditures in support of universities exceed 30 percent of government R&D budgets. Between 1991 and 2003 the trend was for this percentage to increase. There is no equivalent national number for the US because most universities are either private or funded by one of the fifty states.

---

[39] One major developed country recently moved away from this model. Japan's prestigious national universities have been spun out of the government as independent educational corporations. They must apply for government funding based on merit. They are also being encouraged to become more open to collaboration with business and industry through joint projects and the licensing of intellectual property.

**Fiscal Year: 2002**

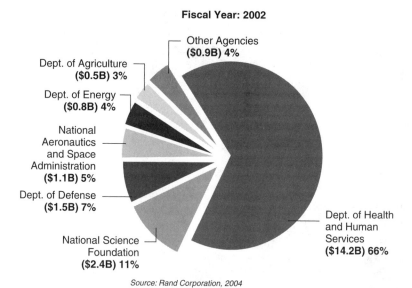

Other Agencies
($0.9B) 4%

Dept. of Agriculture
($0.5B) 3%

Dept. of Energy
($0.8B) 4%

National
Aeronautics
and Space
Administration
($1.1B) 5%

Dept. of Defense
($1.5B) 7%

National Science
Foundation
($2.4B) 11%

Dept. of Health
and Human
Services
($14.2B) 66%

*Source: Rand Corporation, 2004*

**4.8.** Funds provided by US federal agencies to universities and colleges for conduct of R&D in fiscal year 2002. Copyright © 2004 by Rand Corporation. Reproduced with permission of Rand Corporation via Copyright Clearance Center (ref. 40).

Yet the figures for US government funding of universities are actually more meaningful than for those of non-US institutions. The funds tabulated in Table 4.3 support the institutions as a whole. How much goes toward actual research is not broken out. By contrast, data on federal support for research in US universities is available.

In 2002 this amounted to a total of $21.4 billion, rising from $12.8 billion in 1991 for an overall increase of 45.7 percent in constant 1996 dollars. As shown in Figure 4.8, a very large share of the funding, 66 percent, comes from the Health and Human Services side of the federal government and funds research in that sector, primarily in medical schools. The remaining funds ($6.98 billion) come from various agencies and are devoted to scientific and electronic programs.[40]

---

[40] Figure here is drawn from D. Fossum, L. S. Painter, E. Eiseman, E. Ettedgui, and D. M. Adamson, *Vital assets: Federal investment in research and development at the nation's universities and colleges* (Santa Monica, CA: RAND Corporation, 2004), p. 14.

*Table 4.4 The world's top universities[a] and annual R&D funding from US government for US institutions[b]*

| World's top university rankings | World's top universities | US federal funds received in 2002 – excluding medical school ($m) |
|---|---|---|
| 1 | Harvard University | 193 |
| 2 | Stanford University | 178 |
| 3 | University of Cambridge | NA |
| 4 | University of California, Berkeley | 224 |
| 5 | Massachusetts Institute of Technology | 381 |
| 6 | California Institute of Technology | 64 |
| 7 | Princeton University | 80 |
| 8 | University of Oxford | NA |
| 9 | Columbia University | 129 |
| 10 | University of Chicago | 48 |
| 11 | Yale University | 67 |
| 12 | Cornell University | 149 |
| 13 | University of California, San Diego | 223 |
| 14 | Tokyo University | NA |
| 15 | University of Pennsylvania | 118 |
| 16 | University of California, Los Angeles | 193 |
| 17 | University of California, San Francisco | 71 |
| 18 | University of Wisconsin, Madison | 253 |
| 19 | University of Michigan, Ann Arbor | 216 |
| 20 | University of Washington, Seattle | 267 |

Notes:

[a] Ranked by a mixture of indicators of academic and research performance, including Nobel Prizes and articles in respected publications, reported in *The Economist*, "The brains business" (September 10, 2005), 4 (ref. 41).

[b] D. Fossum, L. S. Painter, E. Eiseman, E. Ettedgui, and D. M. Adamson, *Vital assets: Federal investment in research and development at the nation's universities and colleges* (Santa Monica, CA: RAND Corporation, 2004), pp. C-13–C-14 (ref. 42).

A survey published by *The Economist* lists what are believed to be the top twenty research universities in the world in 2005. Of these, as shown in Table 4.4,[41,42] seventeen are in the US. The table also shows

41 *The Economist* "The brains business," 4.
42 Fossum, Painter, Eiseman, Ettedgui, and Adamson, *Vital assets*, pp. C-13 and C-14.

government funds for R&D received by these institutions in fiscal year 2002, *excluding* support of medical school research.

When medical school funding is included the level of federal funding is much larger. For example, the University of Michigan (Ann Arbor) received $429 million in 2002 which increased to $517 million in 2003.[43]

Although there are 3,200 colleges and universities in the US, only a few can support large, federally funded research programs. The recipients of such programs are concentrated in a group of 100 universities, which together receive more than 80 percent of the funds.

There is also great disparity within this elite group of 100. The top twenty schools get the lion's share of 45 percent. The 100th recipient, for example, the University of South Carolina, received $32 million compared to MIT's total of $381 million.

It should be easy to quantify the industrial impact of this level of R&D spending, but in fact trying to correlate university research funding with its ultimate commercial value is a futile endeavor. This is partly due to the nature of the research.

Universities increasingly perform applied research in physical sciences, electronic devices, and software – the kind of research that used to be done in the big corporate research laboratories. It is notoriously difficult to transfer the results of such work into practical applications, because the process is so random and inefficient.

How, then, does government-funded work become available to benefit the industrial world? There are a number of mechanisms, including:

- Open publication of research results, a valuable source of information and ideas for commercial companies.

- Placing graduate students in research projects in disciplines of potential commercial interest to give them job training. These people can become productive very quickly in an industrial R&D environment, which clearly benefits local industry.

- Patenting innovations conceived by the faculty at major research universities, and licensing these patents to industry. As shown in Figure 4.9,[44] revenues to US universities from the licensing of patents

[43] J. V. Lombardi, E. D. Capaldi, K. R. Reeves, and D. S. Mirka, *The top American research universities, 2005*, an annual report from the Lombardi Program on measuring university performance (The Center, University of Florida, 2005), p. 20.
[44] S. A. Merrill, R. C. Levin, and M. B. Myers (eds.), *A patent system for the 21st century* (Washington, DC: National Academies Press, 2004), p. 34; derived from

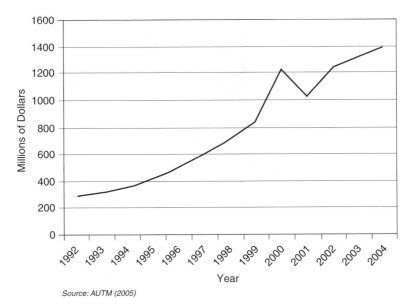

Source: AUTM (2005)

**4.9.** Net license income of US research universities. Reprinted with permission from *AUTM US licensing survey summary: FY 2004* (Northbrook, IL: Association of University Technology Managers, 2005) (ref. 44).

and technology rose from under $200 million in 1991 to about $1.38 billion in 2004.

• Allowing part-time consulting by faculty, usually up to a day a week. This is an efficient way to transfer knowledge to industry while providing financial rewards to the faculty.

• Transfer of new ideas and concepts as seeds for new products, or even for new businesses started with venture capital funding. Some universities allow faculty to take a leave of absence for starting businesses or other commercial pursuits.

• Leveraging of university-developed technologies into new products by existing businesses. We will discuss the cases of Sensors Unlimited and Analog Devices later.

Can one measure the economic value of university research? Unfortunately, no. There is no question that universities such as MIT, the California Institute of Technology, University of Michigan, Stanford University, the University of California (Berkeley),

*AUTM US licensing survey summary: FY 2004* (Northbrook, IL: Association of University Technology Managers, 2005).

Cambridge University, and Carnegie Mellon University have been the source of many important industrial innovations. However, these only gained commercial importance when the concepts were turned into products outside of the university environment.

## Creating clusters of innovation

There is one clear indication of how important a university can be in the development of a technology. It is the increasing influence of clusters of high-tech industries that have grown up around large research universities.

The most successful examples, the ones that everyone else strives to replicate, were built in the San Francisco Bay area around Stanford University and the University of California–Berkeley. Large high-tech industry clusters were also built around MIT and Harvard University in the Boston area, and around Carnegie Mellon University in Pittsburgh.

Cambridge University in the UK has attracted corporate research centers because of the local availability of outstanding talent. It has also generated companies from research in biotechnology, silicon chips, displays, and software. For example, Sir Richard Friend, Cavendish Professor of Physics at Cambridge, is the co-founder of two companies based on his research. Professor Andy Hopper is the co-founder of twelve companies.

The conditions required to create such innovation clusters are not easy to duplicate. They include universities with outstanding faculties, students, and researchers from around the world; funding for research programs from industry and other sources; and the availability of local entrepreneurs and venture capital. The faculty must also be tempted to get involved in new ventures, which usually means offering financial incentives.

The hardest requirement to meet may be the availability of entrepreneurs. This subject is discussed in Chapter 5. Entrepreneurship is not a universal urge among technologists. It is fairly common in the US, Canada, and Israel (and increasingly the UK). But for cultural and regulatory reasons it is much rarer in other countries.

Cultures in which failure condemns the individual to a life of shame are not conducive to entrepreneurism. Unwillingness to assume career risks on a personal level is a major deterrent to starting a new business. To avoid failure, the inclination of technically trained people in those cultures is toward employment in big companies or the government.

Restrictive government regulations designed to protect employment also discourage entrepreneurship in many countries. While it is easy to hire people, the regulations make job elimination very difficult. Since a start-up must be able to trim expenses (including employees) very quickly if it is to survive unexpected financial setbacks, these constraints threaten its existence.

This is a particular problem in some European countries. Jacques Delpla, economic advisor to the interior minister of France, has remarked that "a Google or a Microsoft could not happen in France."[45]

Another problem in moving university innovations into the market is the scarcity of venture capital needed for seed financing of risky business ideas. Funds dedicated to this kind of investing are rare internationally because professional investors favor lower-risk situations where technology issues are well under control. Lam Kong-Peng, director of Singapore's Biomedical Research Council, has commented on the need "to educate Asian investors, who tend to be more cautious than western investors."[46]

## University/industry collaboration: Effective innovation partnerships

The most effective way to exploit university research in building a high-tech industry is to make the two sides partners. For a long time that wasn't even possible, since research in the physical sciences at US universities was a rather primitive affair.

Professors focused on theoretical research because they didn't have adequate labs for more practical investigations. Scientists interested in more practical research in fields like atomic energy or advanced computing gravitated to the National Laboratories, attracted by their excellent facilities and freedom from day-to-day funding pressures.

All that changed in the 1960s with a massive increase in funding for university research from the US Department of Defense. This was a reaction to the surprise launch of two Sputnik satellites by the Soviet Union in 1957. In an effort to close a perceived technology gap with the

---

[45] Quoted in J. Gapper, "Better to be a foreigner than an entrepreneur," *Financial Times* (October 6, 2005), 15.

[46] Quoted in C. Cookson, "Innovative Asia: How spending on research and development is opening the way to a new sphere of influence," *Financial Times* (June 9, 2005), 15.

Soviets, government made funding available to major universities to support research on materials, electronic devices, and computer science.

This was the first time government funding had been provided to facilitate "practical" experimental work. It helped turn the US research university into a quasi-industrial business, ready to take on development programs formerly restricted to corporate laboratories.

Coincidentally, this seismic shift happened just as large US corporations were beginning to de-emphasize basic research in their own labs. As a result of the combination of circumstances, PhD level engineers and scientists were increasingly drawn to academic careers.

Eventually a new generation of researchers, composed of top scientists drawn from the big corporate laboratories, joined university faculties and obtained research grants. They were willing and able to tackle problems addressing industrial and defense needs. This effectively changed the academic attitude toward "practical" research.

For example, Stanford University became a leading semiconductor device research center in the 1960s when a highly-respected scientist, Dr. Gerald Pearson, joined the faculty after a long career at Bell Laboratories. Students trained at Stanford University during that time later staffed many of the Silicon Valley semiconductor companies.

Having started university centers of expertise with government grants, professors soon added consulting to their portfolios of activities. In this way they obtained corporate sponsorship for their research.

The passage of the Bayh-Dole Act in 1980 further encouraged, and in many cases enabled, technology transfers between university labs and commercial companies. The Act allows American universities to retain rights to technology they develop under federal funding. It also permits small or emerging companies to build their business plans on technology developed at research universities.

The Bayh-Dole Act set the stage for risk-taking by entrepreneurs in high-tech collaborations between university labs and venture start-ups. The following two case histories show how different these collaborations could be from one another.

### Princeton and Sensors Unlimited: New company launch

One successful technology transfer involved a novel imaging device operating with non-visible infrared (IR) light. The technology behind the device was developed at Princeton University, and then transferred to Sensors Unlimited, also in Princeton, NJ.

The imaging device has become a key element in night vision systems and various industrial applications. Its basic technology was developed by Prof. Stephen Forrest, a distinguished scientist who had left Bell Laboratories in 1985 to join the University of Southern California, and later relocated to Princeton.[47]

In 1990, Dr. Marshall Cohen and Dr. Greg Olsen, the latter a former RCA Laboratories scientist and founder of Sensors Unlimited, approached Prof. Forrest with a proposal. They wanted to collaborate on the development of commercial imaging systems based on Forrest's invention.

This was not a simple undertaking. The fabrication of such devices is an extremely capital-intensive process. It requires several million dollars worth of equipment and clean room facilities to create even the most rudimentary capabilities for device manufacture.

Fortunately, the university had just won a large grant from the State of New Jersey for the express purpose of moving photonic technologies (including IR detector arrays) from academic laboratories into the marketplace. This was part of a plan to transform the basis of the state's economy from manufacturing to high-tech.

In fact, Prof. Forrest had been recruited to Princeton to manage that state grant through the university's newly-founded Center for Photonics and Optoelectronic Materials (POEM). So a deal was made.

Both Prof. Forrest and Dr. Olsen were sensitive to the different and sometimes conflicting objectives of the university and the new company. The company's goal was to build a profitable business manufacturing new products based on proprietary technology. The university, on the other hand, had as its mission the generation of new knowledge and the education of the next generation of scientists and engineers.

To meet the needs of both parties, all of the initial work, done jointly by Prof. Forrest's students and employees of Sensors Unlimited, was devoted to engineering research: investigating new ideas for making the devices more efficient, reliable, or versatile. This allowed the students to publish papers on new results – a necessary step in meeting requirements for their advanced degrees in electrical engineering.

---

[47] H. Kressel conversation with Professor S. Forrest, July 2005.

At first glance this would not appear to move the company toward its goal of profitability. In fact, the mechanism for financing this initial work did advance this goal.

The joint university/industry team was funded through the Small Business Innovation Research (SBIR) program sponsored by several US government agencies. To get an SBIR grant, a company must win competitive awards for developing new product technologies. A team that took innovative ideas developed at a university, turned them into products, and had an eventual commercial outlet through an aggressive small business was precisely the type of organization the SBIR program was established to support.

After three years of collaborative research, Sensors Unlimited was able to market a unique product that outperformed its competitors. With this product it established customer relationships that gave it strong prospects for future success. At this juncture the company could afford to build its own fabrication facility and begin large-scale commercial production.

Although the initial phase of the relationship had served its purpose, Sensors Unlimited and Princeton University continued to collaborate closely on many projects over the ensuing seven years. The early experiences of working together, of having students collaborate closely with company technical staff on numerous innovative projects, created a culture at both Princeton and Sensors Unlimited that encouraged innovation and teamwork.

This culture served both entities extremely well. A number of Princeton graduates assumed jobs and even significant leadership roles within the company. Sensors Unlimited became one of the world's leading manufacturers of IR detectors.

In 2000 this self-financed company, built on the concept of partnering in innovation with a university, was sold to Finisar, Inc. for $600 million. Two years later the company was reacquired in a management buyback. It is now part of the Optical and Space Systems Division of the Goodrich Corporation.

### Berkeley, MIT, and Analog Devices: New product launch

Sensors Unlimited was a start-up. The industrial partner in this second case history is a large, established company collaborating with not one, but two universities.

The University of California–Berkeley and the Massachusetts Institute of Technology had received complementary Department of Defense funding in sensor technology. Ray Stata, co-founder of Analog Devices, a leading semiconductor company, remembers that faculty enthusiasm for corporate sponsorship of university research dramatically increased as government grants become more difficult to obtain in the 1980s.

Analog Devices was enthusiastic too. It is a billion-dollar manufacturing company that supports product development, but it does not have a central research laboratory of its own. Academic research offered a way to fill that void at modest cost. For that reason the company funded several small research programs at top universities in the 1980s to build on the results of Department of Defense-funded programs in sensors.

The Berkeley/MIT program involved the development of inertial, pressure, and acoustic sensors, all based on a novel semiconductor technology that integrated mechanical and electronic functions into a single device. These devices are used in automobile applications to provide the trigger mechanism for air bags.

This program represents a perfect application of the ABC model of innovation shown in Figure 3.1, with the universities contributing to the concept and early development (the "A" phase of the program), and Analog Devices engineers taking the product from A+ to B.[48]

The result was a huge commercial success. More than 200 million of these devices have been shipped since their introduction in the 1990s. Mr. Stata attributes his company's current (2005) dominant position in this class of sensors to the collaborative programs conducted with Berkeley and MIT.

## Publications and patents: A shift to Asia, universities

Earlier in this chapter we saw how innovation activity is increasing in China, India, South Korea, and other Asian countries. As we have just seen, universities are also attaining new stature as centers of applied R&D.

---

[48] H. Kressel conversation with R. Stata, June 10, 2005.

One would expect that these major changes would be reflected in two leading indicators of increased scientific activity, publications and patents. This is indeed the case.

To get a sense of which countries of origin appeared more often, I surveyed journals from scientific or engineering societies. In the realm of scientific publications, however, all technical journals are not created equal. For a truer picture of the value of the work being published, I focused on a few highly-respected journals.

These are refereed or peer-reviewed publications, meaning that the articles they contain have been reviewed by anonymous experts who recommend their inclusion based on the importance and validity of the results they describe. I looked at four English-language publications covering materials science, electronic devices, and software that attract submissions from around the world: *Applied Physics Letters* (published by the American Physical Society), *IEEE Transactions on Software Engineering* (published by the IEEE), *Communications of the ACM* (published by the Association for Computing Machinery), and *Electronics Letters* (published by the UK's Institution of Engineering and Technology).

A survey of the geographical origin of papers dealing with the physical sciences in two publications, *Applied Physics Letters* and *Electronics Letters*, is shown in Figure 4.10a. It demonstrates that US contributions have dropped from about 60 per cent in 1988 to 30 per cent in 2004. In that time, contributions from Asia (excluding Japan) have increased from 5 per cent or less to between 25 and 36 per cent.

Figure 4.11a shows that articles in the software-oriented journals (*IEEE Transactions on Software Engineering* and the *Communications of the ACM*) exhibit a similar trend. In 1988 US contributions were in the 75 per cent range for both publications, whereas in 2004 they constituted only 35 per cent of the former. (Figures for the latter remained fairly stable.)

Figure 4.10b also shows that fewer and fewer of the authors are affiliated with corporations. In the physical sciences, contributions from corporate research centers declined from 40 per cent in 1988 to 20 per cent or less in 2004. During the same period, contributions from universities rose from 45 to nearly 70 per cent.

In the software field, Figure 4.11b, university contributions have stayed in the 60–70 per cent range, while corporate contributions have

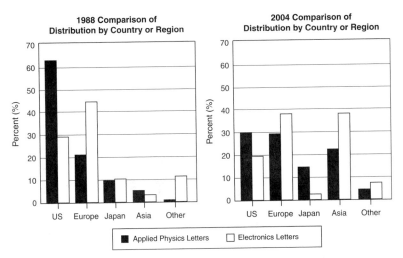

**4.10(a).** The changing sources of publications in the physical sciences and engineering in two publications, *Applied Physics Letters* and *Electronics Letters*, between 1988 and 2004 by country or region of origin.

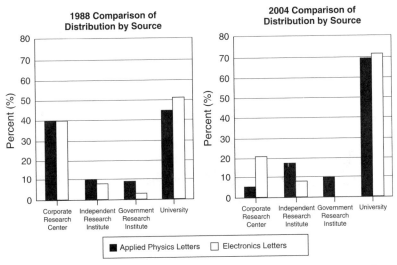

**4.10(b).** The changing sources of publications in the physical sciences and engineering in two publications, *Applied Physics Letters* and *Electronics Letters*, between 1988 and 2004 by the nature of the authors' institution.

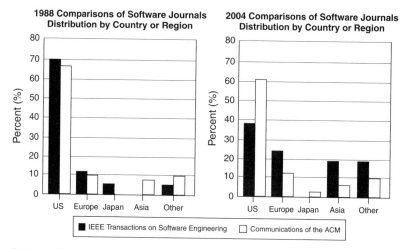

**4.11(a).** The changing sources of publications on software in the *IEEE Transactions on Software Engineering* and *Communications of the ACM* between 1988 and 2004 by country or region of origin.

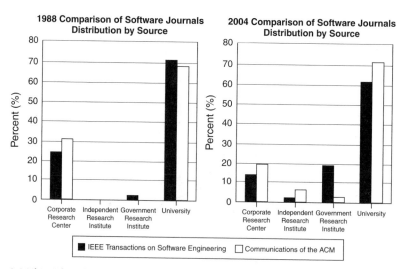

**4.11(b).** The changing sources of publications on software in the *IEEE Transactions on Software Engineering* and *Communications of the ACM* between 1988 and 2004 by the nature of the authors' institution.

declined from the 25–30 percent range to under 20 percent. On the other hand, government research institutes increased their contributions to the 20 percent range in 2004 from a negligible amount in 1988.

Although there are signs of a shift to Asia, the changes seen in the origins of patent filings are not as dramatic. The OECD tracks patents which their owners deem sufficiently important to file in the European Union, the US, and Japan simultaneously.[49] In the year 2000 that number was about 42,000, up from about 29,000 in 1991 (see Table 4.5).

About 34 percent of these patents originated in the US, the same percentage as in 1991. Japan's percentage also hardly changed. About 61 percent of the patents filed came from those two countries, a clear indication of their continuing importance in industrial innovation. China appears as a participant in 2000 after negligible contributions in 1991.

Figures are also available for the number of patents filed in major countries during 2004. These are summarized by Table 4.7.[50]

Japan is in the lead in 2004, followed by the US. Note also the greatly improved ranking of Korea (third) and China (fifth). The table also shows that the Japanese filings were heavily weighted toward the industrial sectors of telecommunications, information technology, and electronics.

It's important to note that the number of patents filed in various countries is not by itself a reliable indicator of their innovative or commercial value. The overwhelming majority of patents represent evolutionary innovations of uncertain potential.

But this activity confirms our observation that Asian countries constitute a growing presence in high-tech innovation. Even if most patents have little individual value, when combined into a portfolio of interconnecting intellectual property they may stake a claim on a larger technical field that is of considerable worth.

Patents are also an indication of long-term strategy. They show that Asian countries embraced innovation as the path to prosperity, and that they have created an environment where research and development are encouraged and supported.

R&D has truly gone global.

---

[49] OECD, *Science, technology and industry*, p. 219.
[50] Table from Thomson Scientific, printed in *Financial Times* (October 12, 2005), 5.

*Table 4.5 Patents simultaneously filed in Europe, the US, and Japan in 1991 and 2000*

| Country | 1991 | | 2000 | |
|---|---|---|---|---|
| | Number of triadic[a] patent families | % of total world triadic[a] patent families | Number of triadic[a] patent families | % of total world triadic[a] patent families |
| United States | 10,217 | 34.1 | 14,985 | 34.3[c,e] |
| Japan | 8,895 | 29.7 | 11,757[c,e] | 26.9[c,e] |
| Germany | 3,676 | 12.3 | 5,777[c,e] | 13.2[c,e] |
| France | 1,783 | 6 | 2,127[c,e] | 4.9[c,e] |
| United Kingdom | 1,250 | 4.2 | 1,794[c,e] | 4.1[c,e] |
| Switzerland | 723 | 2.4 | 753[c,e] | 1.7[c,e] |
| Italy | 659 | 2.2 | 767[c,e] | 1.8[c,e] |
| Netherlands | 568 | 1.9 | 857[c,e] | 2[c,e] |
| Sweden | 391 | 1.3 | 811[c,e] | 1.9[c,e] |
| Canada | 275 | 0.9 | 519[c,e] | 1.2[c,e] |
| Finland | 161 | 0.5 | 489[c,e] | 1.1[c,e] |
| Korea | 93 | 0.3 | 478[c,e] | 1.1[c,e] |
| Israel | 104 | 0.3 | 342[c,e] | 0.8[c,e] |
| China | 12 | 0 | 93[c,e] | 0.2[c,e] |
| Rest of World | 1,116[b] | 3.9 | 2,115[c,e] | 4.8[c,e] |
| TOTAL WHOLE WORLD | 28,807[c] | – | 41,549[c,d,e] | – |

Notes:

[a] Patent filed at the European Patent Office (EPO), the US Patent & Trademark Office (USPTO), and the Japanese Patent Office (JPO)

Year availability

[b] Some 1992 instead of 1991

Standard statistical notes used for science and technology indicators

[c] Estimate

[d] Underestimated or based on underestimated data

[e] Provisional

*Source:* OECD, MSTI database, May 2004, reproduced in OECD, *Science, technology and industry outlook 2004,* © OECD 2004.

*Table 4.6 Number of patents applied for in Japan in 2004 – Compared with five other patent authorities*[a]

| Innovation indicator: Japan compared | | | Patents filed by sector (in Japan) | |
| --- | --- | --- | --- | --- |
| Country | Patents filed | World rank | Sector | Percent |
| Japan | 342,726 | 1 | Telecoms, IT, and Electronics | 43 |
| US | 167,183 | 2 | Chemicals, Materials, and Instrumentation | 37 |
| South Korea | 71,483 | 3 | Automotive and Transport | 10 |
| Germany | 55,478 | 4 | Energy and Power | 4 |
| China | 40,426 | 5 | Pharmaceutical and Medical | 3 |
| Russia | 19,104 | 6 | Food and Agriculture | 3 |

Note:
[a] Where patents are applied for in more than one country, the priority country is the original country of application
Source: Thomson Scientific (ref. 50).

## Innovation: Progress and process

If technology is a life form, innovation is its brains and nervous system. Anywhere a high-tech industry is established, innovation and innovators will eventually appear.

This has certainly proven true for the digital electronics industry as it has become an international enterprise. We still don't fully understand all the implications.

Does the availability of creative, technically skilled innovators in emerging economies pose a threat to the livelihood of engineers in developed countries? Can the innovation process be made more efficient? Can it be made more accountable, to use a popular buzzword? Where will the next wave of innovations come from? How will they be commercialized?

There are no certain answers. However, we can extrapolate from past experience to suggest some possible scenarios.

### The quest for brains

The US has been a technological leader in electronics for as long as the field has existed. It has been fortunate in its ability to attract some of the finest technical minds in the world, domestic and foreign, and offer them an environment for success.

This dominance is now being challenged by the creation of huge pools of technically trained talent in other parts of the world, especially the emerging countries of Asia. The threat is compounded by economic factors, especially cost of labor and shrinking profit margins. First manufacturing moved to areas of the world that offered low-cost production. Now it seems that engineering will join it.

Does this mean that US leadership in electronic technology is certain to vanish as popular fears suggest?[51]

I don't think so, provided that the US recognizes that the competitive arena has changed and develops business strategies that take advantage of the new realities.

The most immediate challenge is outsourcing. The push to reduce costs by accessing resources globally is an unstoppable force. Product development accounts for a large share of the costs in high-tech industries, and there are hundreds of thousands of trained engineers in India and China available to work on product development at lower salaries than their counterparts in more developed countries. The migration of some engineering functions to those resources has already begun, and is certain to continue.

Yet I don't think we are dealing with a zero-sum game where outsourcing will drain developed countries of all their engineering jobs. Mindful of the need to protect their intellectual property, international companies are keeping the more strategic aspects of product development closer to home, or at least in locations where piracy is less of a danger.

There is also the question of quality versus quantity. Large numbers of less-skilled engineers can't equal the value of a few creative minds left free to innovate. At least for now, the true innovators seem to congregate in countries where strategic R&D is being done.

## Measuring the value of R&D

It is an article of popular faith that a higher level of R&D spending translates into greater economic value. Unfortunately, both experience

[51] See, for example, R. Florida, *The flight of the creative class* (New York, NY: HarperBusiness, 2004).

and recently published studies support a contrary conclusion: the correlation between research spending and value produced is poor.

This is certainly true on a national level, where R&D projects are frequently selected to accommodate political constituencies or because they are the pet projects of planners. In the case of corporations, where financial accountability is a watchword, you would expect the correlation to be better. You would be wrong, as some recent research on the subject indicates.

Booz Allen Hamilton surveyed 1,000 publicly-traded companies, looking for the connection between R&D spending and corporate financial performance.[52] The total R&D spending for these companies in 2004 was $384 billion, which on average, represented 4.2 percent of their sales.

R&D expenditures were especially heavy in technology-intensive industries such as software, health products, computing, and electronics. The highest spending rates relative to sales, averaging 12.7 percent, was in the software and Internet sector, while the lowest, 1.5 percent on average, was in chemicals and energy.

The study found that there was no direct correlation between R&D as a function of revenues and either growth rate or overall profitability. However, there was a significant correlation with profitability margins. Companies that spent large sums on R&D were in a better position to get into markets early, and hence enjoy better *gross* profits. (This did not necessarily translate to better net margins once all company expenses were included.)

There are two explanations for the results reported in the Booz Allen Hamilton study. First, translating R&D expenditures into profitable new products is becoming increasingly difficult; second, in a globally distributed industry, comparative spending levels are not meaningful.

The first problem is especially difficult to solve, since companies often have little control over the factors involved. In today's rapidly changing technological landscape, short-term product development projects can miss their market windows, reducing sales and margins. Long-term projects can be made obsolete before they're finished, as

---

[52] B. Jaruzelski, K. Dehoff, and R. Bordia, "The Booz Allen Hamilton global innovation 1000 – Money isn't everything," *Strategy and Business* 41 (Winter 2005), 55–67.

new global competitors emerge from the shadows with cheaper, better products.

The second point goes to the heart of the R&D quandary. It's obvious that in a globally-distributed enterprise the same expenditure level can buy vastly different amounts of manpower. It all depends on how much work is assigned to lower-cost engineering staff in, for example, India.

You could make the case that the international dispersal of R&D leads to a higher level of real investment, resulting in better returns on the research dollar. More people can now be employed for the same budget by shifting significant elements of engineering to locations where manpower costs are lower.

### A new global semiconductor company

New high-tech companies cannot ignore the need to use global resources in order to speed time to market, control costs, and gain access to the best possible talent. The Warburg Pincus investing experience in the past few years has been heavily influenced by these needs, and our portfolio companies are global in their thinking.

As an example, let's look at Raza Microelectronics Inc. (RMI), a company founded by Atiq Raza, which is in the business of designing and marketing very high-performance multi-core microprocessors for use in managing IP networks. Its customers include the world's largest network equipment companies in the US, EU, and Asia.

The company's operations are sited in various international locations. The sites are chosen to fulfill two objectives: finding the best talent in each region, and placing resources as close as possible to the company's customers.

Here is how RMI's talent is distributed regionally:

- **San Francisco Bay headquarters**: Architecture design team, comprised of veterans of high-performance microprocessor design with experience working at major companies in the field.
- **India**: Software engineering, working in close collaboration with the US design team.
- **China**: Software engineering, required for support of specific customer requirements.
- **Taiwan**: Chip manufacturing in the facilities of TSMC, providing world-class, state-of-the-art processes. The US design teams work very closely with the TSMC process engineers to ensure that the

company's products are manufactured for the best possible performance.

- **US:** Prototype, low-volume system design and manufacturing. These are used to allow customers to design RMI's chips into their systems where quick time to market is essential.
- **Marketing and sales:** Worldwide deployment in regions where customers are located.

This business organization has been in place since the company was founded in the early 2000s. It has produced outstanding results in terms of development costs and total time required to produce microprocessors, even though the devices are among the most complex ever manufactured.

## Global innovation and industry

You will note that RMI's microprocessor design team is located in the US – and for good reason. The combination of talent and the many years of experience needed to execute such devices cannot be found elsewhere.

I can repeat this observation about many other fields of activity, including highly-complex software products that some of my portfolio companies have developed and successfully marketed. In hardware and software, there is no substitute for an experienced team, and US technologists have simply been at it longer than anyone else.

So it is not surprising that there is little evidence that revolutionary innovations are as yet originating from the emerging economies. Instead, the newly-established industries are using their low-cost engineering talent to steadily improve productivity and competitiveness.

I expect that companies in these new geographies will eventually gain industry leadership in some sectors. But although this has happened with products like flat panels that are heavily driven by manufacturing technology, it is not as rapid a process as some assume.

We saw how this could be done in the case of ITRI and its role in Taiwan's industrial development. ITRI spurred Taiwan's successful effort to build the leading contract chip production facilities in the world. Taiwanese companies are also leaders in subsystems for computers, including displays. They have become fearsome competitors through relentless attention to reducing costs and improving processes.

South Korea has also been successful in establishing technological industries by blending the best foreign ideas with their own contributions.

Like Taiwan, it developed strong local technical talent while leveraging the availability of large amounts of investment capital from state and foreign corporate sources. Businesses in the two countries used these resources to generate enormous value for their economies.

It was not revolutionary innovation that generated this growth, but evolutionary innovation focused on manufacturing disciplines. There's nothing wrong with that. In any economy, both revolutionary and evolutionary innovation must be supported – and not confused.

The US, by contrast, put revolutionary innovation at the forefront of industrial development. The high-tech industry has thrived in large part because there were entrepreneurs with an appetite for risk, and venture capital organizations willing to help them start new enterprises based on their maverick ideas. New enterprises were born with revolutionary innovations as their business focus.

Many of the best innovators in the US do start-ups. In most countries, the best engineers tend to stay with big employers.[53]

Where does that leave developed economies in face of the new competition? In a world where intellectual property is the core corporate asset but is difficult to protect (see Chapter 7), innovations are of little value unless they can be rapidly exploited.

Globally deployed companies, many of which are headquartered in the developed countries, are best positioned to turn new ideas into valuable revenue generators. The example of RMI is but one among many such new companies that have learned to be global.

I consider the Apple iPod as an example of this process for an established company. Conceived in the US (with locally-developed software) but manufactured in Asia, in a supply chain controlled by Apple, this product is marketed worldwide with huge success. The profits flow largely to Apple and media suppliers in the US.

Innovation is global, but it is not everywhere the same. The trick is to exploit it globally, capitalizing on the differences. In this realm, the developed economies still have the advantage.

[53] Cookson, "Innovative Asia," 11.

# 5 | Financing innovation: Venture capital

In the morning sow your seed and in the
evening do not be idle, for you cannot know
which will succeed.

(Ecclesiastes 9.6)

Just when global competition forced the big vertically-integrated US
corporations to focus their R&D on near-term products, a new support
system for innovation emerged: venture capital.

In the early 1980s, the venture capital industry became a major catalyst
in bringing to market digital technologies that had been incubated
in the corporate laboratories. In the mid-1990s came the rush of venture
investments in start-up Internet companies.

Much as we might deplore the excesses of the Internet bubble, the
many billions of dollars invested there left a permanent legacy of
achievement. It is doubtful that the Internet would have developed as
rapidly as it did, bringing with it the benefits that we enjoy today,
without venture capital. About $200 billion of funding was provided
between 1984 and 2005 to build thousands of new companies.

Having entered the venture capital business in 1983, I had a ringside
seat as these events unfolded. I participated in the successes as well as
the heartbreaks that come with revolutionary periods in history.

In hindsight it looks obvious that the new digital technologies would
find willing investors. That was not the case. In fact, many promising
companies struggled in their early years because the markets were so
ill-defined.

But trailblazers in every era have a hard time finding investors to
fund their exploits. The difficulties Columbus faced in financing his
journeys are well known.

Exploration, after all, is too high-risk for all but the boldest backers.
For that reason, my favorite "angel investors" of the past are two visionary women: Queen Isabella of Spain and Queen Elizabeth I of England.

By supporting Columbus in 1492, Isabella helped open up new horizons for European growth and expansion. Then Elizabeth financed the Sir Francis Drake expedition of 1577 to pillage the very American possessions that Isabella had acquired for Spain. It was an object lesson on how one successful investment enables another: Elizabeth's profit was nearly thirty times her investment!

Inventors and entrepreneurs are explorers, too. They discover new technologies for new markets. Like the explorers of old, they must line up investors to back their enterprises. But finance and marketing do not come naturally to them.

The business talent for promoting an invention and bringing it to the market . . . seems to occur in men in inverse proportion to the talent for creating inventions.[1]

It's no wonder, then, that the relationship between inventors and investors is not always harmonious, even though many of their interests are aligned. The alliance between Edison and the investors who funded the great inventor's electric lighting company in the 1880s is a case in point.

The relationship between the inventor and his bankers, who were to provide the sinews of war for the Edison system, were at best those of an uneasy coalition between opposing interests and temperaments. To the men of capital Edison might be a "genius" and a great "creative force" in himself, but he was still an eccentric, an impetuous character, whose ways of doing business were fantastic.[2]

Managing the relationship between investors and the entrepreneurial inventor is an art form, learned only through experience. Smart investors know that they are not kings and queens who can dictate terms and command obedience in return for patronage. On the other hand, entrepreneurs cannot sail out of sight of their benefactors and follow their own whims.

Investors and entrepreneurs obviously need each other, and unless they work well together there is no way a venture can be successful. Moving innovations into the marketplace requires a flexible, pragmatic

---

[1] Jeremy Bentham quoted in M. Josephson, *Edison, a biography* (New York: History Book Club, 1959), p. 64.
[2] Josephson, *Edison*, p. 247.

approach that addresses the needs and aspirations of investors and entrepreneurs alike.

It also requires patience on both sides. Building a successful business takes years and, as I learned as both inventor and investor, there are no simple formulas for success. Every day brings new challenges and opportunities that must be addressed, and there is never as much time as you would like to make decisions, or enough information on which to base them.

Yet the decisions must be made. To avoid them is to let random events control the future of the business. In the course of this chapter we will see how a number of representative companies responded to this challenge over the years, and either triumphed over changing market conditions or were overtaken by them.

In the digital age the major practitioners of this art have been professional venture capital firms. Since the late 1970s, the availability of multiple billions of dollars of venture capital has revolutionized the way high-tech businesses are created, with the US as a primary beneficiary.

It is no exaggeration to say that venture capital firms have helped create many of the most innovative businesses in the world today. Knowing how this came about, and how venture capital helps entrepreneurs build businesses, is crucial to understanding the technological and economic revolution now in progress around the world.

It is also important to see the larger context within which the venture capital industry operates. We will look at the nature of the entrepreneurs who actually run companies which receive venture funds. In some cases these are the original innovators, but more often they are business executives chosen for their ability to guide start-up companies through the process of growth and success. We will also address the role of angel investors and other funding mechanisms.

## Venture capital: A growth industry

First an innovator comes up with a new product or technology with great commercial potential. The next question is, how do we start a company to bring it to market?

Entrepreneurs of the past typically financed their new businesses with their own money, plus investments from individuals (called "angel investors"), family and friends, and customers. In the US, banks have not been a viable source of capital for start-ups, because federal laws from the 1930s have restricted the ability of US banks to invest in certain classes of assets. Because of this, banks in the US usually back their loans with real assets, not equity. This is not the case in many other countries, where banks are actively involved in financing new businesses.

Although bootstrapping a business without institutional support was the entrepreneur's traditional course of action, it is a very difficult process. Because an undercapitalized company is unlikely to realize its potential, and runs a much higher risk of failure, the entrepreneur must spend a huge amount of time raising funds. Attracting private investors to a company that is not a big, visible financial success is a major challenge, to say the least, and a distraction from running the business.

This is not to say that it is impossible to build a big business without a large influx of outside capital. Microsoft and a few others have shown that it can be done. But this avenue is closed to many entrepreneurs. Most require substantial up-front capital to develop a product, build facilities and gain market attention.

I should point out that there is a middle ground between personal and institutional financing. Entrepreneurs can invest some of their savings, or those of friends and family, to demonstrate that their new product has market potential. Then they can solicit professional investors for serious amounts of capital to grow the business.

Whether entrepreneurs go for institutional backing right off the bat, or wait until they've proven their concept, they will eventually need funds to grow the business. At that point it makes sense to turn to a specialized funding organization that manages a pool of money to finance promising businesses. This structure gives those who invest through the fund the opportunity to get in on the ground floor of a new industry, while providing the entrepreneurs with the means to grow their companies.

From the entrepreneur's point of view, the ideal funding organization would have the resources to cover the company's initial needs. It would also provide the basis for a stable, long-term relationship between investor and entrepreneur. This is particularly important to new companies in rapidly growing markets, where follow-on investments are usually needed to expand a business beyond the start-up stage.

Until relatively recently, however, this approach to funding a new enterprise did not exist in any organized form. Venture capital filled that void.

Venture capital emerged as a modern industry fifty years ago, when professionally-managed limited partnerships were formed in the US in response to the lack of other institutional funding. Their purpose was to finance new businesses that balanced high risk with great potential for growth and profitability.

It is no coincidence that the expansion of venture capital coincided with the emergence of the revolutionary electronic innovations we examined earlier in this book. Large holders of capital, such as pension funds, recognized that they could reap attractive financial returns if they invested in businesses focused on commercializing new technologies and services. These institutions also realized that they needed specialists to manage this investment process, since it was likely to be as complex and fast-moving as the underlying technologies.

As we shall see, over the years the venture capital industry as a whole has delivered superior financial returns to its limited partners (the investors). As always, however, we must keep in mind that industry averages tend to mask substantial differences in how well the individual funds have done.

Venture capital investing is a high-risk, high-reward business where luck, timing, and talent play big roles. Some funds are better managed than others, but many of the companies financed by venture firms turn into financial disappointments or outright failures. The firms with poor returns close their doors when their funds reach maturity, only to be replaced by new firms able to attract a new group of investors willing to bet on yet another new investment team.

## A growth industry

The earliest notable venture capital (VC) company was American Research and Development Corporation, active from 1946 to 1973.[3] Another prominent firm, Greylock Partners, was founded in 1965. It is still active in the industry, having raised $500 million in 2005 for its twelfth fund. Warburg Pincus, where I am a partner, opened its first

---

[3] To avoid redundancy and confusion, for the rest of this chapter we will use "venture capital" to refer to this type of funding. The abbreviation "VC" will be applied to the firms that provide the funding.

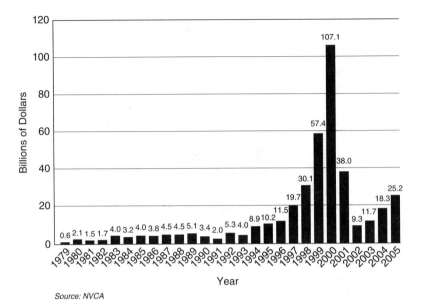

Source: NVCA

**5.1(a).** Capital commitments to US venture funds ($ billions) – 1979–2005. Reprinted with permission from NVCA (ref. 5).

venture fund in 1971 with $41 million of committed capital. In 2005 it raised its ninth fund, for $8 billion.

Recently the industry has expanded at a tremendous rate. The number of venture capital management firms in the US has grown almost tenfold in the last twenty-five years – from 89 in 1980 to 866 in 2005. Their professional staff has mushroomed, too, especially in the last fifteen years, from 1,335 in 1980 to 9,266 in 2005.[4]

Figure 5.1a tracks investments made by VC firms between 1979 and 2005 against the money committed to their venture funds.[5] This demonstrates just how efficient an investment vehicle the venture capital model really is. The annual investments made by venture funds are about equal to the capital committed, as seen in Figure 5.1b.

[4] *National Venture Capital Association (NVCA) Yearbook 2006* (New York: Thomson Financial, 2006), p. 9.

[5] *NVCA Yearbook 2006*, p. 11.

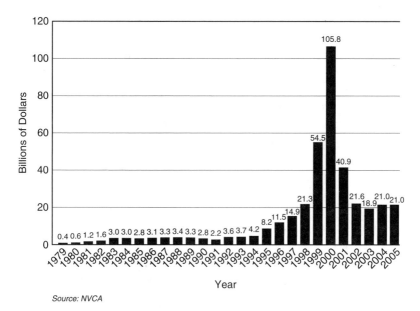

Source: NVCA

**5.1(b).** Investments to portfolio companies (\$ billions) – 1979–2005. Reprinted with permission from NVCA (ref. 5).

## Structure and operation

There is a tendency among the general public to lump all forms of corporate financing together under the general heading of Wall Street. The financial media assume their readers know better, and rarely take the trouble to explain the differences. As a result, readers hear the names Kohlberg Kravis Roberts (KKR), Kleiner Perkins, Loews Corporation, or JP Morgan, and ask, "What's the difference?"

To prevent confusion, we need to highlight some key distinctions that separate one type of investment firm from another. I must caution the reader that the definitions given here are rather arbitrary, as the investment model of a given firm may span a number of strategies. Nevertheless, the distinctions are worth knowing to illustrate the "classic" venture capital investment process.

In general, there are three basic approaches to corporate funding: industrial holdings, private equity, and venture capital. Each is built on a different model, and each is more likely to invest in a company in a particular kind of business at a specific point in its life cycle.

### Industrial holding companies

An industrial holding company owns a substantial percentage of the voting stock in one or (usually) more firms. It uses this ownership to exercise control over management and operations, which it can do by influencing or electing a company's board of directors. The holding company is also called a *parent company*.

There are many well-known holding companies. Berkshire Hathaway, the company headed by Warren Buffet, owns GEICO, Executive Jet, Dairy Queen, and many others. Loews Corporation, mentioned above, is the parent company of Loews Hotels, Lorillard, CNA Financial, Diamond Offshore Drilling, and more.

While industrial holding companies and VC firms both own equity in businesses, there are crucial differences between them.

1. Holding companies generally invest in established businesses, while "classic" VC firms mostly focus on start-ups and early-stage companies.
2. Holding companies can own assets for indefinite periods of time, while venture capital funds have a finite life. They must eventually be returned to the original investors in the funds.

### Private equity firms

Although some large firms have investment strategies that include both approaches, there is a distinct difference between "classic" VC firms and private equity investors. VCs frequently invest in companies that are in the early stages of development. These companies are a risky investment, because neither their technology nor their business plans have been market-tested.

By contrast, private equity firms (also called *buy-out firms*) like to invest in mature businesses that are already profitable. Increasingly, firms such as Warburg Pincus will invest in a broad range of companies with different models, straddling the "VC" and "private equity" model.

Private equity firms use a combination of equity and debt to acquire ownership in their target companies. Big lenders, such as banks, help finance these acquisitions on the basis of the cash flow generated by the acquired businesses.

A typical private equity transaction would be the purchase of a business with a history of steady, multi-million dollar revenues and sustained profitability. The idea, akin to professional real estate investing, is to repay the loans over time with the cash flow of the business, and then sell the business. The strategy of the investors is to focus on improving the profitability of the business while expanding sales, usually with new management.

The realized profit to the investors when they sell the business is the difference between the sale price and the remaining debt. When it works, the rate of return can be very attractive, assuming the interest rates on the loans are relatively low.

Because of favorable market conditions in recent years, this process has been used to acquire larger and larger companies, with the purchase price financed largely by low-interest debt. Acquired companies have included manufacturing businesses, retail store chains, and even software companies. Started in the US, this kind of investing has also become prominent in the EU.

It is estimated that in the years 2005 and 2006, well over $200 billion of private equity capital was available worldwide for equity investment. To date, relatively few private equity buy-outs have involved high-tech companies due to the greater revenue and profitability risks involved, although more of these are being done.

### Venture capital investing

We have already described how VC firms raise funds from limited partners for investment in new or developing companies, and how these investments balance high potential returns against a greater degree of risk.

As VC firms have proliferated, they have increasingly specialized in specific industry sectors, such as communications, software, or biotechnology. They also tend to specialize according to the maturity of the companies they fund, preferring to handle either early-stage companies or ones that have already demonstrated a degree of financial and operational stability.

The specialization reflects their staffing and structure. VC firms are staffed by partners with extensive industrial and financial experience. The general partners who manage the funds have broad discretion in investment decisions.

The amount of funds available varies widely from one firm to another. Some VCs manage as little as $20 million and limit their investments to seeding start-ups. Others, with hundreds of millions of dollars under management, concentrate on investments in more mature companies that need expansion capital.

The compensation structure is reasonably uniform across the industry. VCs receive an annual management fee tied to the size of the funds they invest, as well as a share of the profits realized by the investments.

Because the investments are in privately-held companies, they are illiquid. It normally requires several years of development before a private company's value can be realized through a sale or a public offering of its shares. As a result, the typical life of a VC fund is expected to be at least ten years, at which point the fund is liquidated and its assets are returned to the limited partners.

In general, the median age of a venture capital investment at liquidation is between five and six years.

## How VCs invest

Venture capitalists view their mission as the creation of valuable businesses through active participation in all phases of the development of their portfolio companies. In this discussion, remember that venture investing is a Darwinian process.

That clearly calls for involvement far beyond simply putting investment money into a company. The money is crucial, of course, but VCs must also protect their investments. They have to engage in activities that help give each company its best chance to succeed.

The first step is to select a business for investment. This is not as easy as it sounds, as I discovered when I started my venture capital investing career in 1983. I was deluged with ideas from entrepreneurs, and sorting through them proved very difficult. It's easy to be attracted by exceptionally promising ideas, but many of them never get translated into successful businesses. On the other hand, even average business concepts brilliantly executed often lead to spectacular successes. That's a lesson I learned only over time.

Execution really is crucial. The success or failure of a company can usually be traced to how well its management (and investors) deal with rapidly changing markets and technology. For this reason, VCs also pay close attention to the strategic aspects of the business, and to the people who execute it.

In fact, staying current in a broad range of industries and technologies is an essential requirement for choosing and managing investments. New issues come up almost daily that require managers to make decisions without delay. When they do not have sufficient information, they must rely on their knowledge of the industry.

The best analogy for the process is navigating a small sailboat through perpetually stormy waters, with sharks added for good measure. These circumstances make jumping ship a highly undesirable exit strategy. This is not a business for the faint of heart.

Needless to say, the hazards of the business don't stop VCs from fulfilling their ultimate obligation: developing what is commonly called the "exit strategy." This means finding ways to liquidate their investment and return the proceeds to the firm and its original investors.

I am often asked what VCs actually do in the course of their work. Here are some of the activities that my VC colleagues and I engage in during the life cycle of an investment.

- Identify technologies and/or companies with the promise of creating markets with unusual growth potential.
- Find a management team to build a business around the technology. This is particularly important with firms who collaborate with universities to generate investment ideas.
- Work with entrepreneurs to formulate a business plan based on a perceived market opportunity.
- Negotiate the terms for a new investment.
- Track whether the portfolio company is progressing along the desired lines. If not, make sure that steps are taken to fix the problems, including changes in management.
- Seek opportunities for mergers or acquisitions that can increase the value of a portfolio company.
- Help find sources of capital to supplement the original investment.[6]
- Position the company for an initial public offering or sale.

Obviously, VCs are not and cannot be passive investors. Given the constant challenges of managing a high-tech company in unpredictable and competitive markets, overseeing investments requires constant attention and involvement.

---

[6] It is not unusual to see companies with investments from more than ten different VC firms by the time they go public or are sold. By co-investing with other firms, the original VC can spread the risk while bringing new sources of investment expertise into the venture. The additional funding is usually sought in stages, allowing companies to meet their cash needs throughout their development.

In addition, VCs must have the legal power to make strategic and managerial changes in their portfolio companies in order to carry out their fiduciary responsibilities. The vehicle by which they exercise their prerogatives is the company's board of directors.

The board usually includes representatives from the VC firm, the chief executive officer of the company, and experts selected for their industry knowledge. Outside members play an extremely important role, as they bring a broad range of pertinent experience onto the board.

Typically such boards do not exceed six to eight people, who normally meet monthly for a detailed review of all activities of the business. This structure provides the oversight and control VC firms need to respond quickly to anything that might impact the company's chances of success.

All the attention in the world, however, cannot compensate for an inadequate management team. It takes the very best executives and managers to succeed. While there might be many people hoping to be successful entrepreneurs, only a few prove to be capable of building valuable businesses. So what makes people want to try?

## Incentives for entrepreneurs

Entrepreneurs are notoriously independent-minded. It's reasonable, then, to wonder what attracts outstanding entrepreneurs and management teams to start or join venture capital-backed firms, when they know they will be under constant scrutiny from their investors and their boards.

The answer is twofold: the opportunity to build a new business in a dynamic industry, and the potential to make a meaningful amount of money.

The earnings potential is a powerful incentive. Much of it comes from the fact that a start-up's employees, particularly the professionals, can acquire significant equity in the firm.

The distribution of ownership varies widely from one start-up to another, depending on each company's history and the total amount of money invested from the outside. It is not unusual, however, for employees to own 15 to 25 percent of the economic value of a venture company.

If the company is successful, those employees can share many millions of dollars when the company is sold or has a public offering of its stock. Many of the early employees of Intel became instant millionaires during their company's IPO. This happy example has not been lost on others in the San Francisco Bay area and elsewhere. Many have since attempted to replicate this outcome in their own start-up companies, with varying degrees of success. The more recent IPO of Google, which created a business with a multi-billion dollar valuation, keeps encouraging others.

To make the prospect of entrepreneurship more enticing, a company can command an enormous price in a sale even before it shows any revenues. All it needs is technology valuable enough to attract buyers.

Witness Chromatis, an Israel-based start-up in the network equipment business. Chromatis was acquired by Lucent Technologies in 2001 for stock valued at $4.5 billion at the time. The total venture capital investment in the company was about $50 million. That leaves a lot of room for profit when cashing out.

Such blockbuster returns on investment are relatively rare, of course. Yet despite the odds, many people are willing to take the risk, encouraged by the number of other venture capital-backed start-ups that have created wealth for investors and employees on a smaller but no less welcome scale.

### Options, incentives, and taxes

Employees usually acquire ownership in start-up companies through grants of stock options. While only 3 percent of all US companies offer stock options, they have been part of the employee incentive program in virtually every venture capital-backed company.[7]

Stock options give employees a chance to share in the capital appreciation of a business without incurring any direct cash cost to the company or the employees themselves. In this way they help a company reward talented people while conserving its cash for growth and development.

The practice of issuing stock options has been so successful that many high-tech companies continue it even after they go public. However, in the interest of total disclosure to investors of anything that might affect present or future corporate earnings, the US Department of the Treasury and the Internal Revenue Service have drafted new regulations that force companies to place a value on these options.

---

[7] B. McConnell, "Venture firms fret that a job creation bill could curb the use of stock options," *The Deal* (September 25, 2005), 20.

Valuation of options using these legally required but problematic methodologies has financial reporting implications for both public and private companies. A company offering options to its employees has to record them as an expense. This affects its reported profits for the current year, thereby reducing its perceived value, even though there is no cash transfer to the employee.

## The entrepreneurial society

As venture capital firms proliferate around the globe, it is worth reflecting on the factors that influence people's interest in entrepreneurship.

We should first note that the US has no monopoly on entrepreneurs. People with the talent and drive to create new technology, new businesses, or new industries can be found everywhere.

Still, a large number of foreign entrepreneurs do set up shop in the US. In fact, recent immigrants made up a surprising percentage of the founders of US-based technology companies. One study shows that close to 30 percent of the Silicon Valley companies started between 1984 and 2000 were founded by immigrants.[8]

Immigrants also constitute a large fraction of the engineering staffs at some US technology companies, as we mentioned in Chapter 4. Some will eventually leave to start ventures of their own.

The fact that there are so many foreign-born and, in some cases, foreign-educated entrepreneurs and innovators proves that the entrepreneurial spirit operates across cultures. So-called "national character" is not the decisive factor in promoting or discouraging risk-taking. It is more accurate to speak of "national culture."

Why do these immigrants choose the US as the place to pursue their dreams? The answer seems to be that it offers them a unique set of cultural and business conditions.

First, there is the US attitude toward failure. We noted in Chapter 4 that in Europe and other parts of the world, where examples of successful start-ups are relatively scarce, failure in business carries very unpleasant social consequences. It can brand an individual for life.

---

[8] R. Florida, *The flight of the creative class* (New York: HarperBusiness, 2004), p. 107; derived from A. Saxenian, *Silicon Valley's new immigrant entrepreneurs* (San Francisco: Public Policy Institute of California, 1999).

The fear of failure is a huge deterrent to entrepreneurship in these countries.

In the US, on the other hand, business failure doesn't carry the same social stigma, and it certainly does not end a career. Instead of being shunned by their peers, entrepreneurs with a failure in their past are often given credit for surviving a learning experience. The assumption is that this will help them be more successful in their next venture.

To give one example of this mindset, sociological studies suggest that when people in the UK with the skills to build businesses are presented with entrepreneurial opportunities, they are much less likely to take the leap than their US counterparts.[9]

Entrepreneurship in the US has had another, albeit negative, stimulus: the 1990s breakup of large companies such as AT&T (for the second time), ITT, W. R. Grace, GM/EDS, and Host Marriott.

The turmoil of that era produced a generation of employees who no longer believed that they could gain job security simply by working for a Fortune 500 company. They had seen too many people, including themselves, lose their jobs for reasons unrelated to their skills or performance.

Once these talented employees had decided that lifetime jobs with a single company were a thing of the past, it was much easier to attract them to new, smaller companies that offered exciting work. In essence, they were willing to take larger risks for bigger rewards, because working at established companies was not all that safe anyway.

Europe and Japan have yet to experience a massive restructuring of their own major companies, and that may be one reason why the urge to strike out in a new venture is less common. However, given the entrenched nature of the local culture in most countries, it is unclear whether even the disillusionment that follows such a huge layoff would stimulate entrepreneurship as it did here.

In fact, entrepreneurship is not universal even in the US, in spite of the favorable conditions we have been discussing. Instead, it is a regional phenomenon. For example, entrepreneurs enjoy great local prestige in California, Boston, and other areas where there have been a significant number of successful start-ups. They are looked on as role models for others.

---

[9] See R. Harding, *Global entrepreneurship monitor* (London: London Business School, 2004); and J. Moules, "The start-ups that finish badly," *Financial Times* (November 16, 2005), 9.

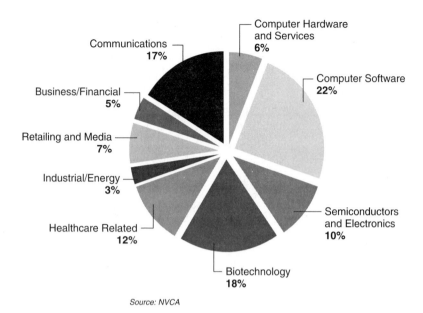

Source: NVCA

**5.2.** Venture capital investments in 2005 – by industry sector. Reprinted with permission from NVCA (ref. 10).

In other parts of the country, where big companies have historically dominated local industry, the situation is quite different. People may talk about entrepreneurship, but they still see it as a less attractive option than joining an existing company for better job security.

### Where venture capital gets invested

Venture capital firms seek out businesses within their investing horizon that have the potential for unusual value creation. These tend to be in dynamic new markets driven by rapid technology changes. Not surprisingly, many venture investments continue to be in the areas of digital electronics, computers, and communications, the original fields that helped establish the VC industry and have delivered sustained returns.

In 2005 a total of $21.9 billion was invested in 2,527 companies. Figure 5.2 shows the distribution of investments by industry sector.[10]

---

[10] *NVCA Yearbook 2006*, p. 12.

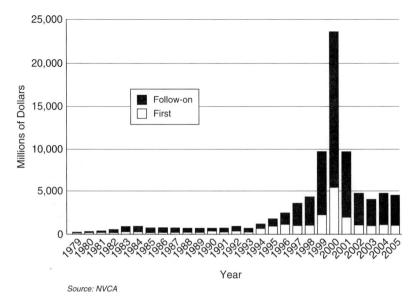

**5.3.** Venture capital investments – first versus follow-on 1979–2005 – software. Reprinted with permission from NVCA (ref. 11).

Semiconductors, electronics, and software together received 32 percent of the money, while 18 percent went into biotechnology and 17 percent into communications.

These investments have fluctuated over the years. As Figure 5.3 shows, software investments grew rapidly between 1995 and 2000. They hit a peak of $23 billion in 2000, then dropped to $4.7 billion in 2005.[11] (The 1999–2001 figures are an anomaly caused by the Internet frenzy.) The money was invested in 729 different companies.

Although the dollar amounts are much lower, investment in semiconductors and electronics follows the same pattern, having peaked in 2000, as shown in Figure 5.4.[12] A total of $2.2 billion was invested in these technologies in 2005.

An amazing example of how quickly venture capital flows into really promising new markets is the investment boom in Internet companies which cut across several of the industrial sectors listed in Figure 5.2.

---

[11] *Ibid.*, p. 48.　　[12] *Ibid.*, p. 51.

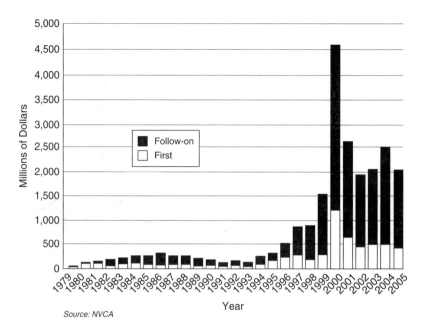

Source: NVCA

**5.4.** Venture capital investments – first versus follow-on 1979–2005 – semiconductors and electronics. Reprinted with permission from NVCA (ref. 12).

Table 5.1 shows the spectacular increase in Internet-related business investments between 1994 and 2005.[13] In the peak year of 2000, VCs invested $78 billion in 4,444 companies.

This represents a staggering 73 per cent of all the venture capital invested that year (a total of $107 billion). In a memorable phrase, Alan Greenspan, then chairman of the Federal Reserve Bank, dubbed such behavior "irrational exuberance." Few of these hastily financed start-ups survived into adulthood. Those that did were pioneers, blazing a trail followed by many others and thereby creating immense value. Some were acquired by the survivors for their technology. Many just disappeared.

## Venture capital growth and profitability

Given the experience of the late 1990s, it is fair to conclude that too many companies are created and financed in hot new product areas.

[13] *NVCA Yearbook 2006*, p. 33.

*Table 5.1 Internet-related investments by year 1994 to 2005*

| Year | Number of companies | Millions of dollars |
|------|---------------------|---------------------|
| 1994 | 151 | 679.06 |
| 1995 | 356 | 1,711.28 |
| 1996 | 694 | 3,736.87 |
| 1997 | 945 | 5,814.86 |
| 1998 | 1,390 | 11,248.96 |
| 1999 | 2,948 | 40,809.49 |
| 2000 | 4,444 | 77,927.14 |
| 2001 | 2,245 | 24,911.41 |
| 2002 | 1,296 | 10,236.52 |
| 2003 | 1,027 | 7,902.08 |
| 2004 | 957 | 7,992.65 |
| 2005 | 914 | 7,545.60 |
| TOTAL | **17,367** | **200,515.92** |

*Source: NVCA Capital Yearbook 2006 (ref. 13).*

I will discuss this subject below. Nevertheless, the venture capital industry has produced attractive returns and continues to attract new capital.

The amount of venture capital available has grown enormously since the late 1970s. As we saw in Figure 5.1a, what started as a trickle of money in 1979 ($600 million) became a flood in 2005 ($25.2 billion). That's an impressive increase even without factoring in the huge peak of $107.1 billion in 2000, during the Internet bubble.

The annual investments made by these funds, also shown in Figure 5.1b, grew at a comparable rate. Note again the huge spike in investments made between 1994 and 2000, the "bubble" years, and the falloff in the subsequent years. This phenomenon will also be discussed below.

## Many happy returns

Why did the amount of venture capital increase so rapidly? We only need to look at the spectacular successes of the early investments to find the answer. In the case of Digital Equipment Company, for example, a total of $70,000 invested in 1957 by American Research and Development Corporation returned $355 million to the investors in 1971.

*Table 5.2 Five-year rolling averages of rate of return in capital funds: Venture capital vs. public markets*

| Five-year period ending | Venture capital | S&P 500 | NASDAQ |
|---|---|---|---|
| 1990 | 6.5 | 9.4 | 2.8 |
| 1991 | 8.6 | 11.5 | 10.9 |
| 1992 | 8.7 | 12.0 | 15.4 |
| 1993 | 11.7 | 10.9 | 15.3 |
| 1994 | 13.2 | 5.4 | 10.6 |
| 1995 | 20.1 | 13.3 | 23.0 |
| 1996 | 22.5 | 12.2 | 17.1 |
| 1997 | 26.0 | 17.4 | 18.3 |
| 1998 | 26.7 | 21.4 | 23.1 |
| 1999 | 46.5 | 26.2 | 40.2 |
| 2000 | 48.4 | 16.5 | 18.6 |
| 2001 | 36.9 | 9.2 | 8.6 |
| 2002 | 27.3 | −1.9 | −3.2 |
| 2003 | 25.4 | −2.0 | −1.8 |
| 2004 | −2.3 | −3.8 | −11.8 |
| 2005 | −6.8[a] | −1.1 | −2.2 |

*Note:*
[a] Average of 600 funds. The top performing 150 funds generated a rate of return of 24.4%.
*Source: NVCA Capital Yearbook 2006* (ref. 14).

Returns like this, far surpassing those available from assets such as real estate, bonds, or publicly-traded equities, were sure to make investors more willing to commit capital to the brand-new electronics market.

Investors in venture capital funds have indeed done well over the years. Of course, the industry's average rates of return depend on what period you are considering. But even taking periodic fluctuations into account, the results since 1990 are mostly positive, especially when compared to the public equities markets.

This can be seen in Table 5.2, where a rolling average of five-year increments is used for the calculation.[14] The only period where the US venture capital industry *as a whole* lost money was during the five years

[14] *NVCA Yearbook 2006*, p. 81.

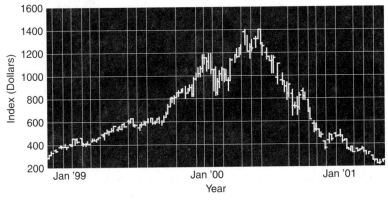

Source: BigCharts.com

5.5. The "Bubble" in the US Telecom Networking Index October 1998 to October 2001. From BigCharts.com quoted by Dr. A. Bergh; reprinted with permission from OIDA (ref. 15).

ending in 2004 and 2005 – a reflection of the excesses in investment of the bubble years. However, it is important to note that the fund performance varies greatly. Of the 600 firms that were surveyed for the results shown in the table for the 2005 period, the *top 150 firms* averaged returns of 24.4 percent for the five-year period ending in 2005. Obviously, the average result for the whole group was weighted down by the very poorly performing funds with bad investments.

Those five years encompassed the "Internet bubble" period, when valuations of publicly held technology companies rose sharply, then suddenly dropped. This is reflected in Figure 5.5, which charts the rise and fall of the public market index for the telecommunications networking sector between 1999 and 2001.[15] This curve is a good proxy for what happened to other high-tech companies in the public market.

You may wonder why we are looking at the valuations of public companies, when VC firms invest in private companies. There is a good reason for this: public company valuations strongly impact the valuation of investments in private firms in similar markets.

[15] Optoelectronics Industry Development Association (OIDA), courtesy of Dr. A. Bergh, 2005; figure from BigCharts.com.

## Public/private correlations

This requires a bit of explanation. Since VC firms must eventually liquidate their investments, they fund a promising business in the expectation of one of two possible outcomes:

- an initial public offering, which offers liquidity to investors and employees by converting their equity into publicly traded shares; or
- the sale of the business for a value in excess of the investment.

If either of these events coincides with a period of huge interest in the company's industry (as was the case in telecommunications and the Internet during the period from 1995 to 2000), a company with interesting technology can receive enormous valuations, regardless of its profitability or even revenues.

For example, during the Internet bubble years, many start-ups in the network equipment business, even those with negligible revenues, were acquired by dominant industry players such as Lucent Technologies and Nortel for prices in excess of $1 billion. The payment usually came in the form of company stock.

The giants paid these amounts because they badly needed certain technologies to compete, and because they could afford the price. At that time Lucent was valued at about $200 billion. In the absence of internally generated products it viewed the acquisition of a start-up with stock (not cash) as the cheapest and quickest way to grow its business.

We have already mentioned Chromatis, acquired by Lucent, as a prime example of the unbelievable run-ups in value that can occur during a bubble period. Chromatis was by no means the only such case. Lucent alone made thirty-eight acquisitions between 1996 and September 2000. Many of the companies it acquired were venture capital financed, and all were acquired at valuations far in excess of their original investments. Most of them were also overpriced. They were benefiting from the bubble mentality of the time, and from the public market valuations in that sector.

Such profitable transactions obviously produced outstanding returns to the investing VC firms. When the public valuation bubble collapsed in 2000, so did the opportunity to sell private companies at outlandishly inflated prices. The effect was predictable: a steep falloff in the profits generated by the VCs and in the rates of return from their funds.

This is not to say that windfall deals are dead. In the preceding chapter we looked at Skype Technologies SA, an Internet voice-over-IP software company. Skype, funded with total VC investments of about $25 million, was sold to eBay in 2005 for more than $4 billion.

According to Thomas Venture Economics and the National Venture Capital Association, of the 107 sales of venture capital-backed companies in 2005, fully 10 percent were sold at prices which represented ten times their total investment. Another 60 percent fetched prices between one and four times their investments, while about 30 percent were sold at prices below the investments.[16]

## The flawed art of picking winners

As we look at these numbers, it is a good time to reflect on the vagaries and risks of venture capital investing. First and foremost, it is a people-driven process. The fund managers must have the savvy to constantly shift and adjust the management of their portfolio businesses when the assumptions they made at the time of investing bump up against new realities.

For example, a common investment thesis is that a new product requirement will emerge some time in the future because of recent government regulations, or new industry standards, or new ways of doing business (like buying products on the Internet). When a VC firm and an entrepreneur base a business on one of these assumptions, they are betting on uncontrollable events. They are investing in new product development to meet a future need that is by no means certain.

If their timing is right, the product emerges just as the need for it is growing. If the start-up hits a big market at the right time, the results can be spectacular. Cisco Systems grew to dominate its market by betting on the emergence of enterprise data networks. Similarly, one of my investments (in collaboration with Joseph Landy), Level One Communications, bet on the emergence of Ethernet networks built with copper wire rather than coaxial cable. The company gained a leading world share of Ethernet connectivity chips in excess of 40 percent before being acquired by Intel in 1999 for $2.2 billion.

---

[16] Reported by C. Murphy and G. White, "Start-up investors stung by weak M&A exits," *The Daily Deal* (October 21, 2005), 2.

Those are happy results. Now suppose that the anticipated market never appears, or competitors get there first with better products, or the market's emergence is delayed by years. In any of these situations the investors have some unattractive options.

They can fund the company to develop a new product for another market, fix the current product to meet competition or, if they believe that the market is delayed, just buckle down, invest more money and keep going. If the investors have lost faith in the management team, which happens when products never work right, they will first try to find a buyer for the business and, if unsuccessful, shut it down.

## Closing a laggard

Faced with a choice between investing more money in a company that's not meeting expectations or walking away, which option does one choose? How do the investors weigh the options and either decide to put up more capital or call it a day? As you can imagine, board meetings where such life and death options are being discussed are very painful, and counsel is likely to be divided.

This extreme situation underlines the importance of a VC firm having awareness of the market and technology, and access to the best talent, in each business sector that it selects for its investments. There are examples of successful companies emerging from early failures. But there are no formulas for success, because no two situations are the same.

It all boils down to the judgment of the decision makers that control the money. At the end, the decision to continue investing or call it a day rests with the venture capital partner responsible for the investment. It can be a very lonely job.

My first experience with such a situation occurred in the mid-1980s. A start-up company called Licom was financed to commercialize the first equipment leveraging then-emerging synchronous network technology. This equipment was built around a new telecommunications industry standard called SYNTRAN, and Licom was first to market.

The product was tested by all of the Regional Bell Operating Companies of the time and worked as expected. The only problem was that production orders never came. Why? Because after SYNTRAN was in place, the industry decided to establish a new standard called SONET, and the big telecommunications equipment companies publicly committed to building products meeting that standard.

There was no way that Licom could develop a SONET product without massive new funding, and the timing for product acceptance was uncertain. I decided to exit the investment. Licom was sold for a very modest sum, which ended up being used to pay off the building lease.

There was a buyer for the company because its engineers were outstanding, so at least some jobs were saved. The net result was a loss of $4 million for my firm, but I never regretted the decision. As it turned out, SONET/SDH products took many years to become accepted, and the market ended up being dominated by the giant vendors.

The lesson is just what you would expect. Technology investing is a very risky business. There is high potential for profitability, but only if you learn when to cut your losses and how to ride your winners.

## The role of the angel investor

VC firms do not have the field of early-stage funding to themselves. Investments in start-ups and early-stage companies by wealthy individuals have continued in parallel with the growth of professionally-managed venture capital funds.

These "angel investors" are frequently organized in informal investment groups that pool capital for specific deals. Angel investors are often the only source of money available to fund new companies at sums under a million dollars. The majority of VCs simply aren't interested in such low funding levels. As VC firms have grown in size, the management of small investments has become impractical.

Yet these small investments add up to a surprisingly large total. A study by the University of New Hampshire's Center for Venture Research indicates that individuals invested $22.5 billion in start-ups in 2004. This exceeds the total amount of venture capital investment for that year.[17]

Given the typically small size of angel investments, this means that tens of thousands of new companies are financed in this way, many of which do not survive for more than a year or two. For those new

[17] R. Sechler, "Angel investors cut start-up funding," *The Wall Street Journal* (November 2, 2005), B2G.

businesses that show promise, however, larger follow-on funding comes in the form of venture capital.

Angel investors provide a valuable service. As seed investors, they help get the highest-risk companies off the ground. These are companies which otherwise would not be funded.

Obviously, angel investors' appetite for risk is affected by the prospect of public market exits, the amount of follow-on capital needed, and their belief in "hot markets" where big profits can be anticipated. Internet start-ups, which usually need only small sums to get off the ground, continue to be favorite investment targets.

Not surprisingly, angel investors became harder to find in times of depressed public market valuations for technology companies.

## Venture capital winners and losers

Right from the earliest days of American Research and Development, the ability to generate handsome returns on capital has been a primary reason for the existence of VC firms. Good returns are possible only if valuable companies are created. In this regard VCs have a large number of remarkable successes to their credit.

They have also had their share of failures, as evidenced by the fact that few start-ups become truly successful businesses. Many of these failures are the result of "herd" investing, which pours too much money into hot new markets. The predictable result is a surfeit of companies chasing the same opportunities.

Let me explain how this happens. In large part it is due to the nature of the business. The way venture capital gets invested is not accidental – money follows emerging technologies that promise to open large new markets.

The innovations trigger a flood of investments into new companies, as entrepreneurs and investors flock to capitalize on their potential. Important regulatory changes that reduce the power of dominant companies can have a similar effect.

We see historical precedents for this behavior in the development of industries before the age of the VC. These include railroad building in the nineteenth century and the emergence of commercial radio and mass-produced automobiles in the 1920s.

The most dramatic example in our time has been the unprecedented flow of capital into Internet-related companies between 1994 and

2000, just as the Internet's potential was becoming widely recognized. Another enormous but short-lived wave of investments was triggered by a landmark regulatory change, the Telecommunications Act of 1996. An extensive discussion of this phenomenon appears later in this chapter.

After a number of start-ups join the race to address new markets, a Darwinian process divides the entrants into three broad categories:

- Early failures that chew up capital, then vanish without a trace.
- Companies that develop valuable products, stay private, and cannot reach the scale needed for long-term viability. These end up being acquired.
- The finalists, a group of companies that rank among the market leaders.

Some of the finalists become consolidators of new markets through successive mergers with their competitors. These companies can reach annual revenues in the multi-billion dollar range.

About 21,000 companies covering a number of industry sectors received initial venture capital between 1980 and 2004. We don't know what happened to them all. However, the record shows that 4,000 of them went public during that time, and 2,783 were sold or merged.[18]

Table 5.3a is a list of venture-backed (and two angel investor-backed) start-up companies that I believe were pioneers in validating major new electronic industry markets. This list and the list in Table 5.3b do not claim to be exhaustive, but they are illustrative of the effect new companies can have.

While these companies did not necessarily invent their core technologies, they were the agents for taking them into the market. Some are still independent and consolidators of their market space; others have been acquired or merged out of existence. All left their mark. We discuss the contributions of some of these examples below.

## Companies that went public

We can certainly count some of the 4,000 companies that had successful initial public offerings (IPOs) of their securities between 1980 and 2004 as being among the biggest winners. Figure 5.6 shows the number of venture-backed companies with IPOs for each year between 1970 and 2005.[19]

---

[18] *NVCA Yearbook 2006*, p. 76.    [19] *Ibid.*, p. 73.

Table 5.3a Selected industry pioneers (venture or angel capital-backed) that became industry leaders (2004)

| Company | Year founded | IPO year | 2004 revenues (millions $) | Product |
|---|---|---|---|---|
| Intel | 1968 | 1971 | 34,209 | Microprocessors |
| Checkpoint Systems | 1969 | 1977 | 779 | Security network software |
| Microsoft[a] | 1975 | 1986 | 38,474 | PC software |
| Apple Computer | 1976 | 1980 | 9,763 | User friendly personal computers |
| Oracle[a] | 1977 | 1986 | 10,557 | Relational databases |
| 3-Com | 1979 | 1984 | 669 | Ethernet networks |
| Seagate | 1979 | 2002 | 6,129 | Disk drives for the masses |
| Sun Microsystems | 1982 | 1986 | 11,230 | UNIX workstations |
| Electronic Arts | 1982 | 1989 | 2,957 | Electronic computer games |
| Autodesk | 1982 | 1985 | 1,234 | Design automation software |
| Cisco Systems | 1984 | 1990 | 23,579 | Computer networks |
| Xilinx | 1984 | 1990 | 1,586 | Programmable logic chips |
| Comverse Technologies | 1984 | 1986 | 959 | Voice mail systems for telecommunication |
| Amazon.com | 1994 | 1997 | 6,921 | Internet store |
| BEA Systems | 1995 | 1997 | 1,012 | Internet transaction software |
| eBay | 1995 | 1998 | 3,271 | Internet auction service |
| Covad Communications | 1996 | 1999 | 429 | Digital Subscriber Loop (DSL) service |
| Google | 1998 | 2004 | 3,189 | Internet search engine |
| NeuStar | 1999 | 2005 | 165 | Telecom inter-carrier services |

Note:
[a] Initial capital from angel investors

Table 5.3b Selected venture capital-backed industry innovators (acquired)

| Company | Year founded | Product | IPO year | Year acquired |
|---|---|---|---|---|
| Digital Equipment Corporation | 1957 | Minicomputers | 1981 | 1998 |
| WordStar | 1978 | Pioneered word processing software for PCs | – | 1994 |
| Ortel | 1980 | Semiconductor lasers for cable systems | 1994 | 2000 |
| Lotus | 1982 | Spreadsheets for PC | 1983 | 1995 |
| Compaq | 1982 | First portable PC | 1983 | 2002 |
| Epitaxx | 1983 | Fiber optic light detectors | – | 1999[a] |
| Level One Communications | 1985 | Ethernet connectivity chips | 1993 | 1999 |
| Maxis | 1987 | Games for PCs | 1995 | 1997 |
| Netscape | 1994 | Internet software | 1995 | 1998 |
| NexGen | 1988 | Intel compatible microprocessors | 1995 | 1996 |

Note:
[a] Acquired by JDS Uniphase

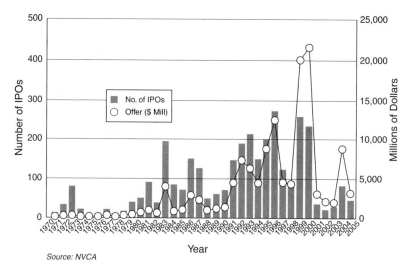

Source: NVCA

**5.6.** Venture-backed IPOs showing number of IPOs and total amount raised. Reprinted with permission from NVCA (ref. 19).

Cyclical peaks occur in years when public markets are unusually receptive to new technology companies. 1996 in particular was a banner year, with 279 companies filing for IPOs. The lean years after 2000 followed the bursting of the Internet bubble in 2000.

Table 5.4 gives us a good overview of the impact of venture-backed start-ups on the public markets.[20] It shows that between 1998 and 2005 these companies represented between 25 percent and 69 percent of all companies filing for IPOs. It is not possible to really do justice to such a large number of businesses. It is interesting, however, to consider some illustrative examples of two classes of outcomes.

If you go back to Tables 5.3a and b, you can see that the venture-backed companies that went public are divided into two groups. Table 5.3a is a selection of companies that pioneered new industrial sectors and, as of 2005, remained leaders in their defined markets. Included are two very prominent companies, Oracle and Microsoft, that were started with angel investment rather than professional venture capital. Note that many are multi-billion revenue businesses.

Table 5.3b lists a sample group of pioneers that have been acquired by other companies. Included are two companies, WordStar and

[20] *Ibid.*, p. 73.

*Table 5.4 Number of venture-backed IPOs vs. all IPOs*

| Year | All IPOs | Venture-backed IPOs | Percentage |
|------|----------|---------------------|------------|
| 1998 | 312 | 79 | 25% |
| 1999 | 477 | 263 | 55% |
| 2000 | 353 | 242 | 69% |
| 2001 | 83 | 37 | 45% |
| 2002 | 77 | 21 | 27% |
| 2003 | 68 | 27 | 40% |
| 2004 | 190 | 83 | 44% |
| 2005 | 183 | 45 | 25% |

*Source: NVCA Capital Yearbook 2006* (ref. 20).

Epitaxx, that deserve mention for their contributions, although they were private when acquired.

One characteristic of market leaders is their ability to assimilate technologies from other sources. While the companies selected in Tables 5.3a and b are notable for their own technical contributions, they were also adept at commercializing innovations developed at universities or in big laboratories. For example, the core idea for Cisco's first product originated at Stanford University. Google's Internet search engine idea also grew out of academic research at Stanford (see Chapter 8).

Another common trait of market leaders is the ability to ensure their futures by acquiring other companies with innovative products, rather than relying totally on internal development. For example, some companies in Table 5.3b ended up being acquired by companies in Table 5.3a.

The fact that companies merge is a reflection of the natural tendency of markets to consolidate around a few strong leaders. As a result, the life span of many new innovative companies as separate entities tends to be only a few years.

## Companies that joined others

The ultimate winners are companies that have had an IPO and successfully maintained their identities and market leadership. But that's not the only way to succeed as a new, innovative organization.

We will now turn our attention to the companies listed as acquired or merged in Table 5.3b, to see how they fit into the innovation pattern. They are representative examples of many other companies that fit this category.

- Digital Equipment Corporation (DEC) pioneered minicomputers. It was acquired by Hewlett-Packard, which also acquired Compaq, the pioneer in small portable personal computers.
- Netscape eventually disappeared into America Online, but not before demonstrating the commercial value of Internet functionality.
- Maxis pioneered the personal computer game market for adults with its SimCity product, one of the best-selling such games in history. It was acquired by Electronic Arts.
- Level One Communications, under the leadership of Dr. Robert Pepper, pioneered Ethernet connectivity chips. It was acquired by Intel in 1999.
- NexGen was the first chip company to demonstrate the ability to generate microprocessors compatible with the Intel operating systems. Headed by Atiq Raza, it was acquired by AMD, which has continued to be a serious competitor to Intel, in part thanks to NexGen's technology.
- Lotus, a leader in desktop personal computer software, was acquired by IBM. It is no longer a significant player in its original market.
- Ortel commercialized semiconductor laser technology developed at the California Institute of Technology and was acquired by Lucent Technologies in 2000.
- Epitaxx, founded by Drs. Greg Olsen and Vladimir Ban, established the commercial market for high-speed detectors for fiber optic communications. It was acquired by JDS Uniphase in 1999.
- WordStar developed early word processing software for PCs, demonstrating the value of such software. It eventually disappeared in a series of mergers.

## Winners and losers: Telecommunications

Among the thousands of venture-backed companies that have disappeared are many firms whose demise was the result of poor timing or plain bad luck, not faulty technology. These unfortunate companies often bet on market opportunities created by government regulations, only to see those opportunities evaporate. While government regulations can create winners, they also create lots of losers.

For example, the Telecommunications Act of 1996 was a landmark event, triggering one of the largest speculative investment bubbles in history. As you would expect, quite a few companies built to exploit this opportunity wound up as victims of its collapse.

The Act effectively opened access to local phone lines owned by the Regional Bell Operating Companies (RBOCs, separated from AT&T under the Consent Decree of 1984). For the first time, non-Bell System companies could offer services directly to consumers and businesses. These new companies were called Competitive Local Exchange Carriers (CLECs).

At the time of the Act, the RBOCs controlled about $100 billion in annual revenues. CLECs were funded by an enormous wave of private and public capital from people seduced by the hope of gaining a piece of that market. The emergence of the Internet only added fuel to the speculative fever. Investors had visions of unlimited demand for communications capacity, and hence unparalleled growth in data traffic revenues.

Many equipment and software vendors soon emerged, offering new products to equip this growing industry. This period coincided with the emergence of new broadband communications technologies such as Digital Subscriber Loop (DSL), which were just reaching the commercial stage.

Covad Communications, to name one participant in the frenzy, was founded in 1996 with Warburg Pincus capital as the first company to capitalize on the emerging opportunity. Unlike most others of its kind, it has successfully remained an independent provider of data services, though not without financial turmoil after the collapse of the Internet bubble in 2000.

Neustar is another successful company spurred into existence by the Telecommunications Act. Neustar, led by Chief Executive Officer Jeffrey Ganek and Chief Technical Officer Mark Foster, is a Warburg Pincus venture capital-backed spin-out from Lockheed Martin Corporation. The company provides the interconnection service that enables local number portability in the telecommunications industry. The industry needed this functionality, since the Act mandated that carriers find a way for telephone subscribers to change service providers without losing their phone number. Neustar went public in 2005.

Unfortunately, the Act spawned far more losers than winners. With so many new companies offering equipment, software, and services to the telecommunications industry to entice investors, there was a herd-like rush of private and public capital into the market, launching a classic speculative boom/bust cycle.

Inevitably, massive business failures followed. It is estimated that more than $2 trillion of public value in publicly traded securities was lost between 2000 and 2003. As mentioned earlier, a good proxy for this cycle is Figure 5.5, which shows the public-market Telecom Networking Index rising to 1,400 in October 1998 and dropping to 200 in October 2001.

To understand what helped fuel the overinvestment spree of the late 1990s, let us scroll back and see how matters looked to investors at that time. Consider two examples: the semiconductor laser industry, selling key components in the deployment of fiber optic communications systems; and new software products that addressed the emerging needs of the carriers.

## Disaster in the electro-optic component business

The opening of the communications market to newcomers in 1996 started a race to build new fiber-optic communications systems in the US and elsewhere. Companies that supplied semiconductor lasers for these systems finally began to look like an attractive investment.

As Figure 5.7 shows, sales of lasers began rising sharply in 1998.[21] Looking at the laser business in 1997, a shrewd investor would have projected rising demand and a resulting industry order backlog. The investor would have valued a laser business opportunity on the basis of a $6 billion market, which it actually reached in 2000.

As evidence for this valuation the investor only had to look at such public companies in the laser business such as JDS Uniphase, with a stock price that reached $145 in 2000. The investor could even have been led to extrapolate industry growth past the $6 billion total, and factor this into the estimate of company value.

As it happened, this demand was not sustainable. It represented the peak of a rapid buildup in capacity. Demand was sure to level off or even decline as fiber-optic systems deployment slowed. At that point there would be a huge amount of overcapacity in the laser industry.

[21] OIDA, courtesy of Dr. A. Bergh, 2005.

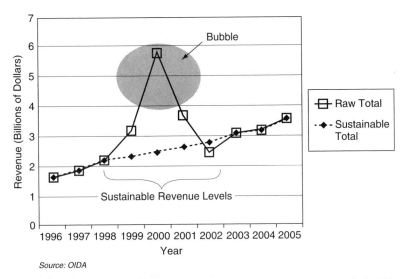

*Source: OIDA*

5.7. Worldwide laser diode sustainable revenues. Removal of bubble (1999–2002) in worldwide laser diode market history to show sustainable revenue levels. Reprinted with permission from OIDA (ref. 21).

Unfortunately, investors assumed that the market saturation point was still a long way off. They believed that the build-out of new systems would be driven indefinitely by the insatiable demand for Internet traffic bandwidth, which was believed to double every month or so.

The market ran out of steam much sooner than they anticipated. As the downturn came, component suppliers saw their revenues drop sharply and their losses rise. That $145 share of JDS Uniphase fell to $3 later in 2000. Built by acquisitions of multiple electro-optic component start-up companies, it was no longer in a position to buy anything.

Private companies supplying lasers were hit hard, echoing what was happening in the public markets. Their values declined sharply as it became clear to investors that only massive infusions of new capital could cover their losses and ensure their survival.

## A new software market emerges

In 1996 it became evident that the advent of the CLECs, which used the existing infrastructure of the RBOCs to reach their customers, was creating a new market for interconnections.

With the dawning of the multi-vendor era, voice and data traffic would have to traverse several carriers to reach its destination. The only way to ensure end-to-end delivery was to create a platform for inter-company communications, so that a CLEC could automatically request service from another CLEC or a big carrier.

For a company like Covad to provide data service over lines owned by the local phone company, for example, it has to send a request to the local company to connect a new customer. Though it may seem unbelievable today, just a few years ago this service request was submitted by fax, and it took a great deal of manual work to implement the service.

An electronic interface between Covad and its line providers to process and implement such requests would obviously be far more efficient. However, this meant developing sophisticated software platforms to interface the internal systems of Covad with those of the phone company.

Here we have a good new product idea. It was clearly a much better solution than the existing methods. No invention was needed – just software development. The target customers were well defined.

Unfortunately, the same idea occurred to several entrepreneurs almost simultaneously. They all began to solicit venture capital. Some won financial backing, because investors believed that interconnection software was part of an emerging $4 billion market serving the liberated telecommunications industry.

It was evident, however, that this market would support only one ultimate winner, because the interconnection platform had to be standardized to broadly serve the industry. The company that acquired the largest customer base in the shortest time would ultimately dominate.

Four new companies, backed with a total of $200 million of venture capital among them, entered the market between 1996 and 2000. Their software was not interoperable. Fierce price competition ensued as each company tried to gain market share and reach profitability. Every sales opportunity became a battlefield, resulting in ever-declining prices for both software and support. By 2002, none had reached critical mass, all had high fixed costs, and all were losing money.

To make the situation worse, CLECs began to go out of business after 2000 and software sales dropped. Management and investors recognized that a merger was needed to build one viable business, but nothing happened. Each competitor expected the others to fail, clearing the field for its product.

At the end, three of the companies were out of business, and the fourth was acquired in 2003 for a modest valuation. There was no last man standing.

## The globalization of venture capital

VC firms typically focus on local companies. An old truism heard in long-established California firms is that they only invest in companies within a one-hour riding distance from their offices.

This has changed. Attractive deal opportunities are now found globally, leading many of the larger US and European firms to start investing internationally.

Warburg Pincus, headquartered in New York City, began developing a global strategy in the 1980s. Today we have offices and partners in Asia and Europe. Smaller VC firms have branched out as well, usually partnering with local firms for their overseas investments. It is now taken for granted that technology firms must have a global outlook, and investors are acting accordingly.

Figure 5.8 shows the venture capital raised by high-tech companies in countries in Asia and Europe from 2002 to 2004. Israel, with a population of about six million, invests more venture capital than China. In 2004 about 55 per cent of the money came from foreign sources.[22] In China approximately 90 per cent of the capital invested in 2004 came from outside the country.[23]

The case of Israel is especially interesting, because it bears upon many of the points we have been considering.[24] We noted in the preceding chapter how actively Israel's government encourages the formation of innovative companies. Venture capital has a huge impact on the Israeli economy as well.

There are many opportunities for venture investments in the economy. One of the reasons for this is Israel's unusually large proportion of highly-trained engineers relative to its total working population. The country has a large percentage of the kind of people who produce innovations. There are 135 engineers per 10,000 employees in Israel, compared to seventy in the US,

---

[22] *Israel Venture Capital Research Center Yearbook* (Tel Aviv 2005), p. 29.

[23] M. Maschek, "China is ripe for VCs with the right approach," *Venture Capital Journal* (November, 2005), 41.

[24] "Punching above its weight: The secret of Israel's success," *The Economist* (November 12, 2005), 64.

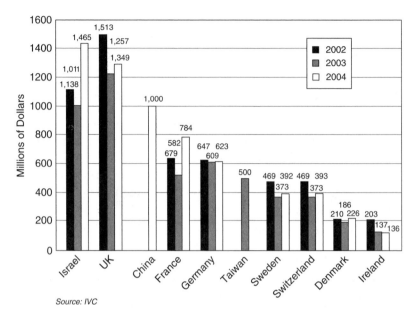

Source: IVC

5.8. Venture capital raised by high-tech companies in various countries (2002–2004). Reprinted with permission from Israel Venture Capital (ref. 22).

thirty in Taiwan, and twenty-five in South Korea. It has proven relatively easy to attract this technical talent into new entrepreneurial companies.

Another plus is the well-developed state of Israel's venture capital industry. It often co-invests with foreign venture capital firms and the venture arms of foreign corporations.

In spite of the global nature of the funding, opportunities to join companies with great growth potential are not only available, they are often home-grown. The success of Israel's own high-tech companies is indicated by the fact that there are over 100 Israeli companies listed on New York stock exchanges, the third highest number in the world after the US and Canada.

Israel's size also predisposes its entrepreneurs toward operating inter-nationally. Because of the small domestic market, every Israeli company is focused on exports. It has become common for high-tech companies to keep engineering development in Israel while locating other corporate activities (including the office of the CEO) outside of Israel to be close to customers, whether they are in Europe, Asia, or the US.

The socioeconomic impact of Israel's new companies is huge. About half of all Israeli exports in 2005 were high-tech products, and many of the participating companies had venture capital funding at some point in their development.

If you want to see how the development and commercialization of innovation can build a country's economy and shape its relationships with the rest of the world, you could do worse than to look at Israel.

## The continuing role of venture capital

There can be no question that companies financed by venture capital have played a central role in transforming the digital innovations of the past thirty years into new products and new markets.

As is always the case in the open competition of a capitalist system, there were winners and losers among the companies created to sell these products. Some went on to market dominance; some validated the new markets but were eventually absorbed by emerging market leaders; and many simply disappeared.

We should not belittle the achievements of the second group, the thousands of firms that were eventually bought out by better-positioned competitors. While their contributions may have been modest and impossible to measure in the aggregate, they were no less real than those of the survivors. It is fair to say that these companies brought new products and technologies with them that enhanced the value of their acquirers.

They also provided R&D that could not have been performed inside the acquiring companies. In their case we can think of the role of venture capital as financing the creation of new products rather than building stand-alone companies. Many venture-backed Israeli companies fit into this category.

Does this investment strategy make economic sense for investors? The answer depends on market demand and the rapidity with which new technologies reach the commercial stage. In fast-moving industry sectors, established companies are frequently handicapped in their internal product development by a lack of specialized talent. Acquiring the talent along with a start-up is cheaper and faster than trying to build a product team from scratch.

The prices paid for start-ups often reflect the value of the talent needed to develop their products as well as the market potential of the products. On this basis, building a company with the intent of selling it rather than establishing a stand-alone entity continues to be a successful model for everyone involved: the VCs and entrepreneurs behind the start-ups as well as the purchasers of the firms.

By funding innovators and creative entrepreneurs, venture capital not only seeds new markets, it also influences or transforms existing markets and the companies that dominate them. We have seen how any new development of major magnitude sets off a wave of investments in direct proportion to the perceived opportunity. This can shake up even industry giants to the point of redirecting their strategies. To take the most salient example in recent history, the success of Netscape forced Microsoft to focus on the Internet. More recently, Google has revolutionized the Internet business model.

A similar transformation is now under way in the software industry. We discussed in Chapter 2 the impact that the Internet is having on the software industry, especially in relation to software as a service (SaaS), a model under which customers access and use software on line as they need it, and pay for only what they use.

This development, now being aggressively marketed by start-ups, threatens to upset the whole commercial structure of the software business. Up to now software vendors have derived the bulk of their revenues from licensed software that sits on the user's computer. There is ample evidence that in 2005 the management of Microsoft began to take this new threat seriously, and that the company will respond.

Without the emergence of VC-funded start-ups, ready to challenge incumbents with new ideas and products, technology markets would be much less exciting places – and the US would lose a great deal of the innovative energy that has been its strength in the past. Creative force, builder of value, or commercial gadfly, venture capital continues to add its own value to the world of innovation.

Before leaving the topic, we need to remember that venture capital financing alone is not the answer to reversing the decline of US manufacturing. The capital committed does not address the massive sums needed for such businesses. Therefore, industrial capital needs to partner with venture capital to build the capabilities that allow major markets to be addressed.

New markets that will be created by process-intensive innovations require long-term capital outlays and making such commitments is outside the usual pattern of venture capital investing. By and large, VC-funded businesses are more likely to be based on services, software, semiconductors, and the design of equipment whose production will be outsourced. However, in partnership with industry, it is possible to envision a successful path to commercialization as long as the manufacturing technologies are sufficiently innovative to overcome the higher labor costs in the US.

# PART III

# Global reach, global repercussions

# 6 | *Manufacturing: Globalizing faster than ever*

M A D E in China. Or in Malaysia, Sri Lanka, Singapore – the list goes on.

Discovering that your clothing was sewn in a newly industrializing Asian country no longer raises eyebrows. But now it seems that even the most advanced electronic products are produced in a developing country, and that the manufacturing job is an endangered species in the US and Europe.

This is an exaggeration, of course. Yet the kernel of truth within it is undeniable: there has been an extraordinary migration of manufacturing, most notably of electronics, out of developed countries since the 1980s. What's more alarming is that the trend shows no signs of slackening.

To a large extent the rapid globalization of manufacturing is a consequence of digital technologies reaching maturity. Technological changes have altered the nature of the competitive advantage derived from manufacturing.

To take just one prominent example, contract manufacturing as a stand-alone business has become a major global activity. Developed in Asia, it represents a total reversal of the historical business structure, under which integrated manufacturing was considered a competitive advantage for practically all product suppliers.

Furthermore, as we discuss later, even software writing has become part of the outsourced "manufacturing" sector, a development made possible by the remarkable advances made in the technology of software development.

This chapter will trace the trajectory of outsourcing through product sectors and whole countries, revealing the pattern of exporting low-skill assembly jobs to foreign shores in search of cost reductions.

As manufacturing plants close in developed countries it has become fashionable to dismiss the importance of manufacturing "hard goods."

There are supposedly more advantages to be derived from creating industries based on software and services.

That's not how I see it. I am convinced that, especially in large economies like the US, a selective but competitive manufacturing sector is a prerequisite for sustaining the innovation that ensures economic prosperity in the first place.

My experience in setting up a new factory, recounted later in this chapter, taught me that the process of innovation extends right onto the production line, particularly in industries where process technologies are important value-creating elements. If you are not well versed in manufacturing, your ability to innovate in these industries will atrophy. Naturally, this presupposes that such industries will be highly innovative to compensate for their greater labor costs.

Whether to manufacture, then, is the wrong question. What to manufacture and what to outsource is a more fruitful line of inquiry for product companies. The answer to this question will vary by industry, and will change with the rapid transitions in global markets, but it must be asked. History shows that it is far from being a new dilemma.

## Early outsourcing

The Industrial Revolution started in England in the mid-eighteenth century. Fifty years later its triumph was assured. The old agricultural and commercial society had given way to the new industrial economy. Machinery had replaced tools in most industries. Factories were superseding craft producers. In the pantheon of economic heroes, commercial entrepreneurs had displaced merchant adventurers.

But clouds were on the horizon. In the 1800s, though Britain was the dominant manufacturing nation, its reign was already being challenged. In the second half of the century major competitors, primarily the US and Germany, emerged with more efficient manufacturing processes that allowed them to commercialize innovations faster. By 1900, it was evident that the industrial landscape had changed. British industry, made complacent by its long supremacy, was finding it difficult to respond to the competitive threat.

The national response set the tone for a debate that has echoed down to our own day. There was a bitter debate between advocates of free trade and those who supported protection to shield local manufacturers from "unfair" foreign competition.

Free traders argued that industry needed to migrate to higher value products. In their view, the new competitors were welcome to the "commodities." Protectionists, fearing the collapse of important industries, strongly disagreed. A tirade by Joseph Chamberlain, who campaigned for Prime Minister in 1905 on a platform of trade protection, will sound all too familiar to observers of more recent trade wars.

Your once great trade in sugar refining is gone; alright, try jam. Your iron trade is going; never mind, you can make mousetraps. The cotton trade is threatened; well what does that matter to you? Suppose you try doll's eyes. But for how long is this to go on? Why on earth are you to suppose that the same process which ruined the sugar refinery will not in the course of time be applied to jam? And when jam is gone? Then you have to find something else. And believe me, that although the industries of this country are very various, you cannot go on watching with indifference the disappearance of your principal industries.[1]

The issues of this debate are as pertinent today as they were 100 years ago. Many developed economies must now make painful adjustments as the manufacturing of more and more products shifts to lower-cost developing countries.

Manufacturing is the vehicle that moves innovations into the market. Its offshoring (to use a newly fashionable word) carries implications that resonate far beyond the simple calculus of comparative labor costs.

It raises such questions as, to what extent is the discipline of production a necessary component of innovation? Does losing dominance in production spell trouble for a country's ability to compete in a field? Does manufacturing still have a place in developed economies? If so, what is it?

Recent trends are beginning to provide the answers to these questions.

## The China factor

If Joseph Chamberlain were speaking today, he would probably single out China as the most prominent usurper of manufacturing jobs. Virtually every international company from Panasonic to Wal-Mart has production facilities or manufacturing partners on the Chinese

---

[1] Quoted in R. Skielelsky, "The Chinese shadow: II," *The New York Review of Books* (December 1, 2005), 32.

mainland, rolling out everything from clothing and Barbie dolls to televisions and notebook computers.

China is not alone, of course. Eastern Europe is increasingly viewed as an attractive location for labor-intensive manufacturing.[2] However, China is the most important and fastest-growing new site for manufacturing electronics, the product area of interest to us. Therefore much of our discussion of the globalization of production will focus there.

The current wave of manufacturing displacement from developed to emerging economies grew out of cost-cutting efforts by large, established companies. This is reflected in the fact that 87 percent of China's exports in 2004 came from plants owned wholly or in part by foreign companies. Chinese customs data puts the share of exports controlled by foreign companies at 60 percent.[3]

These companies believe that they have no choice but to leverage lower labor costs to remain competitive.[4] They can do so because of five major factors that help drive the process. You will note that automation and communications, made possible by rapidly improving computerized and networked digital systems, have played a key role.

1. The growing intelligence and efficiency of computer-enabled manufacturing equipment. Digital controls make it easy to relocate machinery, because the need for expert human operators is reduced.

2. Massive improvements in air, ground, and sea transportation. Sending materials and supplies to offshore factories and returning finished goods to consuming countries is now a matter of days rather than months. That's fast enough to respond to the most volatile markets.

3. High-speed digital communications, which make managing offshore facilities and transmitting design information as efficient as walking next door for a meeting.

4. The liberalization of economic policies, especially in China and India, to encourage foreign investments.

5. Various forms of government subsidies for high-tech industries.

---

[2] This trend has attracted much attention recently. See both "The rise of nearshoring," *The Economist* (December 3, 2005), 65–67; and J. Ewing and G. Edmondson, "The rise of central Europe," *BusinessWeek* (December 12, 2005), 50–56.

[3] Quoted in D. Barboza, "Some assembly needed: China as Asia's factory," *The New York Times* (February 9, 2006), C6.

[4] M. Dickie, "China's challenge changes the rules of the game," *Financial Times* Special Report (October 19, 2005), 1.

Once these enabling mechanisms were in place, the outsourcing of manufacturing happened with remarkable speed. In 1972, when Richard Nixon became the first US president to visit what was then called "Communist China," no one could have imagined that China (and other developing countries) would become major players in electronic manufacturing so quickly. It's as if the Industrial Revolution happened overnight, not just in China, but throughout Asia.

## Electronics and automation

To better understand the pace of global electronics manufacturing, it is worthwhile to define the three major segments of the industry and discuss how value is created in each. We will start with the transformation of materials and end with final product assembly.

- **Process.** The process sector creates value by producing, transforming or combining materials. Some of its products can be used on their own, but most are components designed to be integrated into more complex electronic or optical devices. Because this kind of manufacturing involves the mastery of complex processes, it is built around a great deal of proprietary intellectual property. The manufacture of semiconductor devices falls into the process category.
- **Intermediate assembly.** In this sector, value is created by assembling components into functional assemblies or subsystems. The assemblies are used in turn to build more complex machines. Examples of intermediate assembly include packaging semiconductor devices, assembling electro-mechanical components, and attaching components to printed circuit boards.
- **Final product assembly and test.** At this stage all of the components and subsystems from the process and intermediate assembly sectors are integrated into complete products such as computers, television receivers, or cellular handsets.

Each of these three sectors involves different disciplines. Each also requires different levels of labor input, capital investment, and intellectual property.

The first sector, process-intensive production, creates the most value because it is based on proprietary knowledge. Since the process sector demands a high level of skill and knowledge from its workers, the quality of the labor force is more important than its quantity. This sector also demands substantial capital investment. In all of these respects processing closely parallels materials-based R&D.

Assembly operations, on the other hand, are more labor-intensive than capital-intensive and only semi-skilled workers are required. For this reason the assembly sectors are the first to migrate to countries with low labor costs. Later, when a country has developed a skilled and knowledgeable workforce, it can begin building a process sector.

This pattern has been repeated in country after country across Asia. It started when Japan ramped up its electronic assembly sector in the 1950s. By the 1970s it was building up its process industry, in the form of state-of-the-art semiconductor fabrication plants (fabs), just at the time that the intermediate assembly of electronic components started migrating to Korea, Taiwan, Malaysia, and the Philippines. Korea and Taiwan started building semiconductor fabrication plants of their own in the 1980s.

Now China is creating its own process sector, having gained experience in electronic assembly over the past two decades. Modern semiconductor production began there in the late 1990s.

### From craft to automation: Technology enables migration

Long after Joseph Chamberlain railed against the loss of England's manufacturing dominance, the West is once again watching a manufacturing sector shift overseas, to Asia. This time it's happening faster, powered by computerized production equipment that automates complex production processes.

The principles of automation have been common knowledge since the dawn of the Industrial Revolution. Adam Smith, in his famous 1776 book, *An inquiry into the nature and causes of the wealth of nations*, observed that the essential element in raising manufacturing volume and reducing costs is the division of labor into ever finer specializations. Machines play a central role in this transformation.

For the sake of simplicity Smith chose the example of a pin factory where production was divided into eighteen separate operations. But he stressed that his observations applied to the most complicated production processes.

In every other art and manufacture, the effects of the division of labour are similar to what they are in this trifling one. The division of labour, so far as it

can be introduced, occasions, in every art, a proportionable increase in the productive powers of labour. . . .
Everybody must be sensible how much labour is facilitated and abridged by the application of proper machinery.[5]

Smith's most important point was that the division of labor and use of machines not only increased productivity, it reduced or eliminated the need for skilled craftspeople in production operations.

Modern digital data processing accelerates this changeover. It embeds the highly specialized skills originally required to make electronic products into microchips and the machine tools they control.

With computer-controlled machines carrying out precision operations at high speed, there is no need for large numbers of skilled production workers. A manufacturing plant requires only a few people with the knowledge to program and maintain the sophisticated machinery. All other positions are semi-skilled, and a new workforce can be rapidly trained to fill them. This has made the migration of manufacturing easier than ever.

## Automating transistor production

Automation does not happen overnight. It may come as a surprise to learn that even semiconductor manufacturing was once a hands-on craft. It took forty years, but we have progressed from an industry of skilled craftspeople working within vertically integrated companies to a standardized, highly automated production environment where a turnkey plant can be placed practically anywhere in the world with the appropriate level of support engineering.

I had the good fortune to learn semiconductor manufacturing from the ground up. In 1961 I was put in charge of building a production line for the first RCA transistor factory to make silicon transistors.

Silicon devices were new at the time, and relatively little was known about silicon as a material for the mass production of transistors. The first transistors had been made of germanium, as were all commercial transistors up to that point.

Having designed a new silicon-based transistor (which became the type 2N2102), and having made a few hundred in the laboratory with

---

[5] A. Smith, *An inquiry into the nature and causes of the wealth of nations* (New York: The Modern Library, 1937), pp. 4 and 9.

the help of one technician, I took on the assignment of setting up a production line in Somerville, New Jersey. The plant was already manufacturing germanium transistors with unionized production workers. Part of my task was training the hourly production workers to use the process recipes that I had developed in the laboratory.

Union rules required that operators be specialized, so each was trained in only one part of the process. Cross-training was not permitted. We did this over two shifts, which meant that I sometimes spent as much as fourteen hours a day watching each part of the process as the devices made their way through the evolving production line.

There was little automation, as we used mostly home-built equipment. To make matters more difficult, there was not much overlap with the processes used to make germanium transistors. Finally, the actual assembly of the devices was a laborious job where leads were attached in a machine produced by a local mechanic.

The first batch emerged after a few days. It was a reasonable success. Device yields were about 40 percent comparable to what had been achieved in the laboratory. This was high enough for a commercial product, given that those devices were expected to sell for five dollars each. My management was delighted and a commercial launch was planned.

During the next few days, as subsequent lots of transistors emerged, the yields got progressively worse, until the fifth batch yielded not one good device!

This is when I learned some hard lessons about what pioneering manufacturing really entails. It is a difficult profession, and requires the same level (though not the same kind) of creativity as device design.

Manufacturing is all about discipline: you must understand and control every variable that could impact product quality and cost. To do so requires obsessive attention to detail, since a random event that affects quality or performance in the laboratory inevitably turns into a source of major disaster as volume ramps up.

That's exactly what had happened in my process. As I looked at how the various transistor lots were manufactured, I discovered that my original recipe had been augmented (without documentation) by the specialized operators on the line, who were trying to fill in process steps that had been left undefined.

The reverse also happened. For example, in the laboratory we had performed some operations (such as rinses after chemical etching) under conditions that were, quite by accident, reasonably "clean." The operators on the production floor did not take this precaution. Our recipe did not mention it, so they were not aware of its importance.

It took two months, and many experiments, to identify all process details and find remedies for the problems. Our method was to introduce controlled process changes in selected transistor lots, monitor the impact on device yields, and put in place the procedures that gave us reasonably consistent results. Finally our product was launched.

In the course of this painstaking process I learned to appreciate the value of a motivated labor force. The enthusiasm and intelligence of the production workers proved instrumental in the success of our experiments. They helped us identify the elements that affected transistor yield, and of course they were vital to the implementation of each process change.

On the negative side, I also stumbled into some of the pitfalls of employing unionized factory workers in a dynamic industry. The biggest problems were changing work assignments, and the fact that non-union personnel (including myself) were not allowed to touch production equipment.

Of course, there was no way I could keep my hands off the equipment (including making needed repairs in the middle of the night). As a result, my days in the plant usually concluded with a written grievance from the union formally charging me with a serious breach of contract. This went on for months, and was considered a joke by most of the people involved.

Over time, as production parameters were better understood, mechanization was introduced and some of the human factor removed. However, some of the best ideas for automation in the early days came from factory personnel, and all of the early machines were built at RCA.

The first commercial equipment was designed for automating transistor assembly on metal headers. A company called Kulick and Sofa introduced such a machine, and I believe I was their first customer. They were among the earliest of the many specialized equipment vendors that have supplied production equipment to the electronics industry over the years. Many of these vendors helped define new processes through their innovations.

## The automation of integrated circuit (IC) production

The discrete transistor in a metal can, made on a primitive production line has evolved into the integrated circuit with many millions of transistors on a single chip, produced in a sophisticated IC fab.

If the semiconductor industry had remained in the manufacturing mode of the 1960s, electronics could not have transformed the world. Digital technology could only make its full impact on institutions, businesses, and ordinary consumers when the power of millions of logic gates was condensed into small, manageable circuits.

Manufacturing rose to the challenge. Technical innovations led to the fabrication of devices in smaller and smaller geometries. At the same time, the equipment to produce them steadily increased in speed. With more chips on a wafer and higher-speed production the cost per device dropped dramatically.

In circular fashion the product improved its own manufacturing process. The digital processing revolution launched by these devices (discussed in Chapter 2) gave engineers the tools to design even faster mass production equipment, ensuring ever-decreasing unit costs.

The 2N2102 single transistor is still commercially available. The way it is made bears little resemblance to my production recipes. It is still only one transistor on a chip, but with the transistor count on modern chips reaching one billion, production processes have evolved in a form that was undreamed of in a simpler age.

Highly-automated processing equipment, driven by digital computers running process-defining software, takes the human element out of manufacturing. Humans are not really involved except to maintain, load, and program machines. Devices are automatically tested by digital computers, not by hand as my trained operators and I used to do. The result is that device yields for many semiconductor devices in an automated plant are well over 90 percent.

Given the huge volumes, high yields, and extensive automation in modern fabrication plants, there is very little direct labor cost. The single 2N2102 transistor now sells for pennies.

Of course, the production equipment costs far more than it did forty years ago. My original production line was equipped with machines built for a few thousand dollars; a modern, high-volume, state-of-the-art factory costs billions. But volume justifies the cost, and as the role of automation continues to grow, equipment vendors increasingly add value by making advances of their own in process technology.

## Globalizing IC production

To gain an appreciation for how automation makes it easy to move production lines around the globe, let us look at a single step in the production process: controlling transistor feature dimensions, and thus the physical size of the transistor, on a wafer.

As we have seen, shrinking the transistor is the key to progress in digital electronics. Reducing this process to a repeatable formula at the wafer level took years. Engineers had to define the complex relationships among materials, processes, and device parameters in IC manufacturing.

Only after this was done could computers be programmed to control the hundreds of process steps needed to make a transistor at the desired feature size. Totally new process technologies, capable of controlling materials at the atomic level, were invented to take advantage of this capability.

Much of the early work on manufacturing innovations was driven by the requirements of the US Department of Defense (DoD), which needed advanced devices for space and military systems. For example, Very High Speed Integrated Circuits (VHSICs), a DoD-funded program in the early 1980s, focused on the then-challenging reduction of feature sizes to 1.25 micron (1,250 nm).

One process innovation that produced much smaller feature sizes over the years involved delineating some of the thin film materials that are coated on the wafers. The dimensions of the lines thus formed have to be very accurately controlled. This innovation eventually displaced processes that called for the use of a chemical etch.

Chemical etch processes were always troublesome. Controlling them required tight control of the chemical bath, its temperature, and the characteristics of the material being etched. Production workers learned to adjust their parameters to achieve the desired dimensions only after years of experience. As feature sizes continued to shrink, chemical etching became more and more impractical.

The innovation that eventually replaced chemical etching involves using heavy atoms to blast atoms of the thin film right off its surface. This approach uses the atomic-level computer controls mentioned above to consistently produce devices at the proper feature size. Two versions of the technology exist: reactive ion etching (RIE), also called plasma etching, and sputtering, which is RIE without the ions.

The processes are conducted inside a vacuum chamber. Computer-controlled sensors inside the machine monitor the thickness and composition of the films being etched and stop the process at the appropriate time. The machine can be adjusted to perform this operation on different films with process recipes provided by the equipment manufacturer. In principle, the equipment is no more difficult to use than a washing machine. But it does cost a lot more. Prices start at about $1 million per unit.

Over the years plasma etching and sputtering systems have produced a drastic decline in direct labor costs for electronic device manufacturing. For example, when Zilog replaced its mid-1980s 6-inch wafer factory with an 8-inch wafer facility in the early 1990s, the direct labor force was reduced by a factor of three, while device output doubled.

A similar transformation has occurred in the production of opto-electronic devices. The early semiconductor lasers of the 1960s were produced at RCA Laboratories in the simple handmade apparatus shown in Figure 6.1.[6] The temperature in the furnace where the chemical processes occurred was adjusted manually, using the temperature indicator on the left.

Automation has replaced this relatively crude device with the equipment shown in Figure 6.2, which costs about $4 million.[7] Its chemical technology is more advanced than that used in the 1960s process, and the chamber carries out all process sequences on its own, using software-driven recipes.

At this level of automation, the production of even complex components can quickly proliferate around the world. It is estimated that there are dozens of machines in Taiwan used to make semiconductor lasers for videodisc players. The final products are assembled using low-cost labor in China. The light-emitting components are produced for just a few cents each.

[6] H. Nelson, "Epitaxial growth from the liquid state and its application to the fabrication of tunnel and laser diodes," *RCA Review* 24 (1963), 603–615. Photo courtesy A. Magoun, David Sarnoff Library.
[7] Photo courtesy Sarnoff Corporation.

**6.1.** Early growth apparatus for liquid phase epitaxy of semiconductor lasers developed at RCA Laboratories in the early 1960s. From Nelson, "Epitaxial growth." Photo courtesy A. Magoun, David Sarnoff Library (ref. 6).

## Automation as an industry

What happened in semiconductor production has occurred in other fields as well. The enormous advances in computer-aided automation have transformed process industries such as plastics, steel, and chemicals. But the most profound impact has been felt in the electronics industry, where automation affects every sector from process to final assembly.

For Japanese companies, automation quickly came to represent not just a tool, but a business opportunity. Taiwan, Korea, and other Asian countries were already emulating Japan's electronics industry, using efficient, large-volume manufacturing of electronic products as an entry point into high-tech mass markets. But Japan went beyond manufacturing and assembly to establish a position in the tools that made automation possible: the equipment used in electronics manufacturing.

From the viewpoint of Japanese industry, this was a logical next step. Japan had long dedicated itself to achieving superior manufacturing efficiency as a key competitive advantage over Western suppliers. Because of this commitment, Japanese companies have been leaders in manufacturing innovation for decades. By investing in the production equipment business they helped preserve their lead in manufacturing technology while opening up new markets for their products.

**6.2.** Automated fabrication equipment used to manufacture semiconductor lasers. Photo courtesy Sarnoff Corporation (ref. 7).

### Circuit boards to wristwatches

The initiative started with specialized tool vendors serving local industry. Soon these companies were competing against established vendors in the world market.

For example, the first commercial machines for automatic assembly of components on printed circuit boards (including pick-and-place and wave solder equipment) were produced by US companies. Today, however, Fuji, a Japanese company, is a leading vendor of equipment used to assemble the printed circuit boards for computers and similar products.

Other major Japanese suppliers of automated equipment for electronic manufacturing include Canon, Hitachi, Nikon, Olympus, and Shinkawa. The roster of names represents leading companies in the fields of optical technology, instrumentation, electrical machinery, robotics, and computers. This reflects the diverse expertise these companies bring to the field of electronic manufacturing.

Japanese companies have been highly successful in establishing a dominant position in electronics production equipment. To cite just one example, they have essentially taken over the market for supplying equipment to makers of wristwatches.

They were up against formidable competition from Switzerland, a country whose famous watchmakers had dominated wristwatch manufacturing virtually from its inception. The Swiss watch industry had spawned Swiss machine tool manufacturers who offered highly specialized wristwatch assembly equipment.

However, as the mass market electro-mechanical watch industry moved to Japan, the automation equipment business went with it. In 2004, four Japanese companies accounted for 85 percent of the approximately $1 billion annual market for assembly machines.

In fact, Citizen, one of the largest watchmakers in the world, began focusing on automation to reduce the cost of watches early in the transition from mechanical to electronic timepieces. Today a few Citizen workers can produce the same number of watches as hundreds did only a few years ago.[8]

## Redefining the sources of competitive advantage

In the bygone age of vertical integration, US companies took pride in operating their own manufacturing facilities. Having your own

---

[8] P. Marsh, "Watch and learn to outperform the Swiss," *Financial Times* (May 1, 2005), 11.

production plant was seen as a competitive differentiator. Once manufacturing techniques were standardized, however, this viewpoint
changed.

By standardized I mean that the industry adopted a set of common
physical forms for component parts such as semiconductor devices,
resistors, capacitors, and connectors, so that they could be mounted
onto the circuit boards (also standardized) that all electronic equipment uses. Standardization made manufacturing easier and less labor-
intensive, since it was feasible to design assembly machinery to handle
the bulk of the work.

As the complexity of these boards escalated, however, the machines
became proportionately more expensive, making assembly an increasingly capital-intensive process. The only way to drive down assembly
costs was by increasing the scale of production.

It became evident that companies selling computers, for example,
could reduce costs by shifting the production of their boards to specialist companies. Since they served the needs of multiple clients, their
volume was high enough to fully utilize the assembly equipment. The
combination of high-volume production and full utilization meant
lower cost was virtually guaranteed.

## Subcontracting process and assembly

Over time the manufacturing process evolved to the point where contract manufacturers took responsibility for every aspect of production.
They started with their customers' designs and ended with the delivery
of a fully-tested end product.

This trend has permeated the industry. In the process sector, while
very large semiconductor companies such as Intel can justify their own
chip fabrication plants, smaller ones are choosing to use contract fabs.
We will look more closely at these "fabless semiconductor companies"
in more detail below.

In the assembly sector, even equipment companies are increasingly
outsourcing the manufacture of final products. For example, Avaya
Communications, a global leader in the enterprise telephone equipment business, sees its competitive advantage residing in its innovative
product design, proprietary software, targeted marketing, and customer service. Not only is its equipment production outsourced, but

much of its standards-based software products work with products from other vendors, not just Avaya.

If the belief that vertical integration gives a company competitive advantage has not completely vanished, it is no longer a central article of faith, even for large companies.

## Outsourced assembly: Subsidiaries and contractors

The same drama has played out on the global stage. As Western companies began to seek access to low-cost foreign labor, they had two choices: build their own facilities, or contract out activities (primarily component assembly) that required only semi-skilled labor to local companies in the various countries.

The largest companies usually chose to build their own plants. Visitors to Padang in Malaysia, an early low-cost production destination, cannot miss the vast expanse of manufacturing plants carrying the logos of the best-known companies in the world, such as Texas Instruments.

Smaller companies that could not justify the cost of building standalone facilities chose to use specialized contract manufacturers. The implications of this decision deserve a closer look.

## The advantages of contract manufacturing

The economic advantage of using contract manufacturing specialists is clear: large-scale production, the model for the contract industry, drives costs down. But there is another reason to take this approach to the final product: the chance to tap the expertise of manufacturing experts.

As I learned from starting production of the 2N2102 transistor, manufacturing is a difficult discipline to master. It is not a robot-driven process, even with the most sophisticated equipment. While industry standards increasingly shape product specifications, and automation handles many of the routine aspects of manufacturing, there are still many areas where manufacturing specialists can be the source of further cost reduction by improving equipment utilization and modifying a process for higher efficiency.

Many of the cost reductions grow out of the "design for manufacturing" discipline. For this to happen, product designers must have a full appreciation of all cost and quality elements at the earliest stages of product development. This may sound obvious, but in fact it is extremely difficult to achieve. There are now specialist manufacturing companies that make "design for manufacturing" their value contribution.

## Types of contract manufacturers

Contract suppliers come in all sizes, capabilities, specialties, and ranges of service offerings. Most fit into one or another of the following three categories.

### Contract assembly

Among the earliest specialists in the electronics manufacturing business were companies that assembled components on printed circuit boards. Flextronics, a company which operates globally and had about $16 billion in 2004 revenues, is a prime example of this approach. Many of its factories were purchased from equipment manufacturers who decided to outsource their production to Flextronics.

### Original Design Manufacturers (ODMs) – Turnkey manufacturers

Over the past decade a new breed of contract manufacturers has emerged in Taiwan. Called Original Design Manufacturers (ODMs), these companies provide more than a single manufacturing function. Although they do not own the branded electronics products they manufacture, they participate in the design of the products and take responsibility for their production.

In effect, they perform all product development and production functions except the creation of the original concept and the marketing and sales of the finished item. For example, ODMs are completely responsible for manufacturing cellular phone handsets for some major international brands. This is a large business: the top ten ODMs in Taiwan had combined revenues of $49 billion in 2004.

Among the leaders in the turnkey manufacturing arena is Hon Hai Precision Industry Ltd. (trading as FoxConn Corporation). FoxConn was founded in Taiwan in 1974 by Terry Gou to make plastic switches for television sets. Today it is one of the world's largest contract manufacturers of computers, consumer electronics and other electronic products for Apple, Cisco, Dell, Nokia, and other well-known brand-name companies.

FoxConn is widely known for its expertise in low-cost, high-speed production of all kinds of high-quality mechanical/electronic products. It starts with a client's concept for a product, then takes responsibility for sourcing parts, molding plastic elements, assembling the product, and testing it.

FoxConn's contribution is a turnkey system that completely removes the burden of manufacturing from the product owners. At the end of the process FoxConn delivers a complete product, ready to ship. The market and product risk remains with the customer.

The secret of FoxConn's success is fanatical attention to detail and continuous improvement in cost and quality, plus an extraordinary ability to manage a complex supply chain where some components are produced internally and others, like chips, are purchased.

The company makes maximum use of capital equipment by continually improving its utilization. Factory workers are trained in this discipline, and are at the forefront of the effort. In one case, after many months of tweaking, the company was able to reduce molding time on a part from two hours to thirty minutes. It sounds trivial, but an organization that knows how to focus on details every day emerges as the low-cost, highest quality vendor in the long term.

FoxConn's success has been dramatic, as indicated by its size. Of its total of 160,000 employees, about 100,000 are in China. The company has a large technical staff, including 15,000 mechanical engineers and 15,000 electronic engineers who work with customers and on the manufacturing floor. The central R&D organization in Taiwan numbers 5,000 people. The company's revenues were about $16 billion in 2004. Generous stock option plans motivate management and professionals.

This is a new breed of company, born of the specialized modern discipline of electronic products manufacturing.

### Independent semiconductor fabrication plants (fabs)

Contract manufacturing, as we saw above, has extended its reach beyond assembly to semiconductor processing as well. Large, sophisticated fabrication plants now produce ICs for many clients who don't have their own fabs – and even for some who do.

Independent contract fabs owe their existence to the standardization of most production processes. This allows a single facility to implement many chip designs from almost anyone, anywhere.

The economics of IC production are also a significant factor in driving business to the contract fabs, especially from smaller clients. As production plant costs have escalated beyond the reach of any company with less than $5 billion in annual revenues, contract fabs have become the only viable route to a product for all but the largest firms.

If the semiconductor industry helped the contract fabs get under way through standardization, the fabs have returned the favor by stimulating industry growth. Starting in the late 1980s, access to fabrication plants offering state-of-the-art processing prompted the founding of many new venture capital-backed "fabless" semiconductor companies.

There are thirty-six publicly-traded fabless companies in the US, and forty in Taiwan. Many others are privately held.

Fabless companies consider the designing and marketing of specialized products as their competitive advantage, and leave manufacturing to the experts. For example, Xilinx, the pioneer in the programmable logic business, relies solely on outside contract manufacturing.

The world's largest and most successful independent semiconductor contract manufacturer is Taiwan Semiconductor Manufacturing Company (TSMC). Founded in 1987, it has an annual production capacity of 4.3 million wafers. Its revenues ($8 billion in 2004) represent some 50 percent of the global foundry market.

TSMC is large and profitable enough to maintain a technical development capability, which allows the company to introduce state-of-the-art processes in a timely fashion. Furthermore, because of the excellence of its management, it offers such outstanding service that companies with their own fabs use it as a supplementary source of supply. This company has proven that manufacturing *as a business* can be highly profitable, even in the most difficult technology on the planet.

## Manufacturing success: The China challenge

We noted earlier the unprecedented rate at which manufacturing has migrated to Asian countries. There are many elements to the story, but none of this would have happened without the decisions made by the Chinese and Indian governments to open their countries to foreign investment in plants and facilities, and to build modern infrastructures to support these plants.[9] To clarify how these incentives worked, we will focus on China, which has had the greatest success in attracting manufacturing.

The emergence of China would not have been possible without massive foreign investment and local subsidies. Between 1983 and 2004 this infusion of money rose from about $500 million to nearly $60 billion per year.

The Chinese government did its share to encourage the foreigners. It offered subsidies they could use to leverage their investments. The government also supported the movement of labor from the country into industrial regions, and the free movement of materials and products in and out of the country. In addition, it lured plants through the rapid development of infrastructure in the coastal regions. Without massive investments in ground, air, and data communications, the commercialization program would not have succeeded.

To ensure that China was acquiring advanced technology, the government mandated joint ventures between foreign corporations and either local companies or Chinese government entities. These joint ventures enjoy a special status, entitling them to financial incentives that further reduce the cost of products manufactured in China.

The joint venture between Alcatel and the Chinese government that led to the formation of Alcatel Shanghai Bell, now a leading producer and exporter of communications equipment, was covered in Chapter 4. The political and financial aspects of China's modernization have been well documented elsewhere.[10]

But a country's great leap forward into modern electronic manufacturing hinges on more than foreign investment and industrial policy. It also requires fundamental changes in the manufacturing sector, supported by upgrades in infrastructure.

To address this need, the Chinese government made massive improvements in infrastructure and business conditions in selected

---

[9] See A. Waldman, "India accelerating: Mile by mile, India paves a smoother road to its future," *The New York Times* (December 4, 2005), 1, 32–33; and M. J. Enright, E. E. Scott, and K. Chang, *The greater Pearl River delta and the rise of China* (Hoboken, NJ: John Wiley & Sons, 2005).

[10] A comprehensive analysis can be found in T. C. Fishman, *China Inc.: How the rise of the next superpower challenges America and the world* (New York: Scribner, 2005).

parts of the country, making it possible for modern plants to operate efficiently and tap into global markets. Their goals included:

- Physical improvements, including the rapid buildup of a sophisticated digital communications network, plus the creation of better roads, ports, airports, and industrial parks.
- The availability of a large number of trained engineers.
- A huge, disciplined labor force, willing and able to staff modern factories at wages far below those in developed economies.

This is not to imply that the government tried to modernize the whole country at once. Like all developing countries seeking to grow their manufacturing sectors, China deployed most of its new infrastructure around industrial parks or "zones" to attract industry. A nationwide build-out would have been impractical, though in fact there has been progress on a countrywide level.

In the manufacturing sector, however, China's rapid advance is testimony to its success in meeting the three goals listed above. A brief examination of its progress will show the extent of its achievement.

## Communications infrastructure

Manufacturing for the world market requires close management of supply chains that may stretch around the globe. Modern information technology, based on reliable, high-speed digital voice and data communications, is vital to this endeavor. It is the nervous system of industry, and a necessary precondition for companies (and countries) that plan to compete in the global economy.

There is ample proof that Asia in general, and China in particular, place high priority on creating a modern communications infrastructure. The speed and extent of the region's buildup of digital communications is clear from its deployment of fiber-optic cable.

Figure 6.3 shows how much high-quality, low-loss fiber-optic cable, used in long distance communications, has been installed in various geographies since 1998.[11] Note the sharp rise in fiber optic deployment in only two geographies: North America (US and Canada) and the Asia-Pacific region. The latter includes China and other rapidly

[11] *Worldwide optical fiber and cable markets: Market developments and forecast* (Nashua, NH: KMI Research, Pennwell Corporation, March 2005), p. 143.

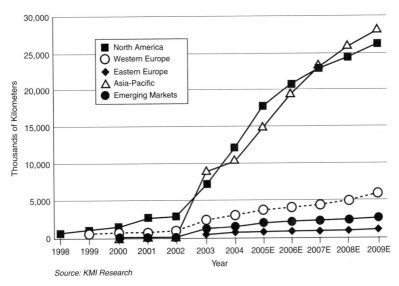

*Source: KMI Research*

**6.3.** Low absorption optical fiber demand by region (1998–2009). Values beyond 2004 are estimated. Figure courtesy KMI Research, www.kmiresearch.com (ref. 11).

industrializing Asian countries. The chart predicts that Asia-Pacific will outpace all other regions in fiber installations by 2009.

Another indicator of the rapid spread of digital communications is the penetration of cell phone service. China now has the largest cellular subscriber base in the world, exceeding 400 million in 2006 and still growing.

## Trained engineering workforce

As we observed in Chapter 4, both India and China annually graduate legions of university-trained engineers. At 600,000 per year, however, China produces nearly twice as many as India, and over five times what the US can claim.

While the standards of training in developing countries are variable, there can be no doubt that China has an abundant pool of technically trained people who can support modern manufacturing at a much lower cost than in the developed economies (see Table 4.1).

## Diligent, motivated production workers

The manpower for the industrial transformation of Europe, Japan, Taiwan, and Korea came from the migration of agricultural workers into factories, in an effort to improve their standard of living. As that flow of labor diminished, and industries had to compete for workers, the cost of production labor increased. The growth of social benefits helped move production workers into the middle class.

In addition, the workers were frequently represented by labor unions with the clout to demand better wages and benefits. The result has been a steady rise in the cost of labor over the past forty years. This was balanced in large part by continuous improvements in manufacturing productivity, the result of successful automation.

This equilibrium has been shattered by the massive entry of new industrial workers in China and other Asian countries into the labor market. It is estimated that about 100 million workers are engaged in manufacturing in mainland China, and that another 200 million could move into the industrial workforce from the countryside if jobs were available.

As one would expect, the wages for Asia's newly-minted manufacturing employees are much lower than for comparable workers elsewhere. While wages for semi-skilled labor are likely to rise over time, employees whose skills are in demand will continue to command higher compensation.

Table 6.1 shows representative hourly labor costs (including plant overhead costs and personal benefits) for semi-skilled workers in typical industrial manufacturing plants in various countries.[12] Estimates of fully-burdened hourly labor rates for China vary widely, but they all peg the pay rates very low, ranging from $1 to $4 depending on skill levels and location.

Obviously, rates this low have unpleasant implications for the manufacturing sector in countries with high labor costs. It calls into question the very ability of these countries to maintain labor-intensive manufacturing as a viable industry without drastic wage reductions.

---

[12] Table created from private communications from various multinational companies. For data on China labor, see www.iht.com/articles/2005/10/20/business/dvd.php (October 20, 2005) (accessed on May 8, 2006).

*Table 6.1 Fully-burdened hourly labor costs (2005).*

| Country | Rate/Hour ($) |
|---|---|
| China[a] | 1–4 |
| Mexico | 10 |
| Malaysia | 11 |
| Hungary | 24 |
| Korea | 33 |
| United States | 51 |
| Japan | 75 |
| Holland | 80 |

Note:
[a] The wide range is dependent on labor skills and locations in China.
Source: Data from private communications (ref. 12).

## Labor vs. IP: Manufacturing in developed economies

Conventional wisdom says that companies – or countries – with higher labor costs will soon find their manufactured goods priced out of the market. This is not always true. Let us look at two examples to test the truth of this assumption.

### High direct labor content

First we will consider a company that produces electro-mechanical components (such as switches) and larger sub-assemblies for consumer products, such as washing machines. The company makes these products in thousands of different form factors, which are assembled in small lots to meet rapidly changing customer needs. The basic products themselves have changed very little over the past few years.

While the metal, electronic, and plastic components of the products are made on highly-automated equipment, their assembly into the customized configurations is a mostly manual process. It is simply impractical to design automated assembly equipment flexible enough to handle such small production lots. As a result, the products have a high percentage of direct labor content.

In 2002 production moved from a factory in the eastern US to China. The Chinese plant employs about 3,000 people, and has equipment

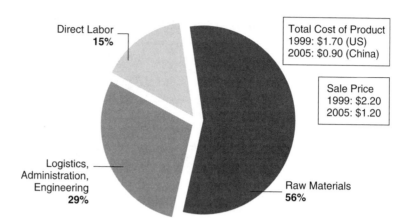

**6.4.** Product cost distribution in China – 2005 compared to the US in 1999. Example of product manufacturing cost of identical electro-mechanical components in China (from private communications/survey, ref. 13).

comparable to that used in the US. The assembly of the many variants of the basic products continues to be done by hand.

Figure 6.4 shows the cost breakdown of the product as produced in China in 2005.[13] The average part costs $0.90 to manufacture. Of this total, 56 percent is for raw materials, while only 15 percent is direct labor. Production engineering and other overhead costs make up the remaining 29 percent.

By comparison, in 1999 the part cost $1.70 to make in the US, even though material costs were the same. The difference in production cost can be attributed largely to higher direct and indirect labor costs. As a result of the cost reduction, the selling price of the product is now $1.20, or about 45 percent lower than the 1999 price of $2.20.

Once a major player in an industry cuts costs like this, the competition has no choice but to follow suit, invent a revolutionary new way to produce the product cheaper, or leave the market. This experience is being repeated in many other industries with products that have similarly high labor content.

---

[13] Chart created from private communications/survey by author.

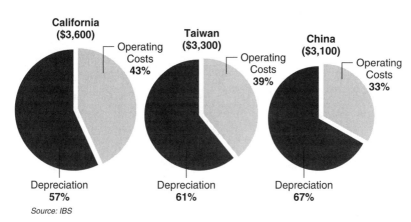

**Sale Value of Product: $6,000 – $15,000**

Source: IBS

**6.5.** Estimated manufacturing cost of 300 mm CMOS wafer in various geographies in 2005 (from Jones, IBS, ref. 14).

## High IP content

Now let's look at a *highly* automated industry with low direct labor content: state-of-the-art semiconductor production. Our test case will be a 300 mm silicon CMOS wafer with state-of-the-art geometries. We will compare the cost of producing it at plants in three different parts of the world: California, Taiwan, and China.

These plants have similar capacities, cost about the same to equip ($2 billion), and use identical commercial equipment. Their processes are highly automated. There is very little direct labor, but a lot of highly-skilled production support engineering. As a result of differences in levels of experience, the productivity of the plants may vary. However, the processes and materials they use are industry-standard.

As Figure 6.5 shows, the cost to produce a wafer in California is $3,600.[14] The comparative cost in Taiwan is $3,300, while in China it is $3,100. Because of the huge investment required for automated production equipment, the dominant costs in all three locations are depreciation-related.

[14] Chart created from data provided by H. Jones, IBS (July 2005), private communications.

Here is where the cost of capital comes into the equation. If money to build these plants is available through state subsidy or at very low interest rates in one particular country, it becomes a key element in site selection, but not necessarily the deciding factor.

Labor cost alone is also not the deciding factor. There is no denying that the lower cost of labor in Asia has an impact on the total cost of production. But as we shall see, there are other important considerations in determining whether manufacturing in California is a viable option.

It is apparent that China offers a substantially lower cost structure than the US in semiconductor production, but there is a big difference between the IC industry and the assembly of components for consumer appliances. For one thing, because there is a high degree of automation in semiconductor production, direct labor costs have much less impact on the final price.

The real differentiators are the value of the intellectual property (IP) embedded in the individual semiconductor devices on each wafer, and the necessity of protecting it.

To put it another way, the production cost of the wafer is less important than the market price of the chips and the ability to rapidly implement new proprietary process technologies to maintain a competitive advantage. Once a wafer has been processed, the devices it contains are cut apart into chips, packaged, and tested. The individual sale prices of these chips vary widely depending on what they are, which is largely a function of how much IP they represent.

Commodity memory chips, for example, are cheapest in terms of value per unit area of silicon wafer, whereas proprietary programmable logic and high-end microprocessors are the most expensive. The commercial value of a given wafer can range from a low of $8,000 for commodity ICs to a high of $15,000 for a batch of microprocessors with high IP content.

In other words, as long as a product is proprietary and its IP translates into high sales prices, modest differences in wafer costs are not a major contributor to the profitability of the company marketing the end products. The ability to come out first with new products and the resulting high value-added content in the products readily outweighs any modest disadvantage in raw manufacturing costs. Under these conditions it is perfectly economical for companies to continue manufacturing products in developed countries, where the most productive technical talent is currently found.

There are also strategic reasons to do so. First, keeping manufacturing on shore maintains close proximity to the best engineering talent, which the manufacturer can tap to rapidly introduce process improvements. Process improvements still drive value if they are quickly translated into better products.

Second, and most compelling, is the fact that it is much easier to protect intellectual property in one's home country than it is offshore. Contracting the manufacture of an IC to an overseas supplier is as much about security as it is about economics.

## Reversing the tide of migration

If proof was needed that IP content trumps labor costs in deciding where to put an automated manufacturing operation, the past few years have provided it. Several new high-tech facilities, including four new semiconductor fabs, are under construction or in operation in the US and Japan, two countries with high labor costs.

In 2002, Toshiba Corporation announced that it had "decided in principle to construct advanced semiconductor production facilities employing 300-millimeter diameter wafers" at two of its major facilities in Oita and Mie prefectures, Japan. The production lines were for system chips for broadband networking and for FLASH memories, respectively.[15] The Oita line, now in operation, is expected to move to the more advanced 45 nm process technology in the future.

According to its press release, Toshiba's rationale for locating these facilities in Japan, rather than in a country with lower labor costs, was to "remain a driving force in technological innovation." In other words, they decided to produce high-IP-content ICs at home, where they could safeguard the IP and further develop the process while reaping good profits – an exact parallel to the strategic position we have been discussing.

Other Japanese companies were announcing new domestic plants at around the same time. Canon, to name just one, started construction on a 29,000 square meter digital camera plant, also in Oita. According to

---

[15] From a corporate press release, "Toshiba to build 300-millimeter wafer semiconductor plants," www.toshiba.co.jp/about/press/2002_12/pr1301.htm (December 13, 2002) (accessed on February 10, 2006).

analysts, the focus on domestic capacity reflected new attitudes toward offshore sourcing of manufacturing, as well as economic reality.

Why the shift? On one level it's because Japanese companies have already reaped most of the potential gains from moving production overseas ... Most of the plants being built in Japan are capital-intensive operations where labor represents a small fraction of the total cost of the finished products ... Japanese companies are also finding that it pays to have researchers work closely with – and in tight proximity to – production teams. Canon ... has slashed the time it takes to develop and produce new digital cameras by a third ... Then there's the question of security. Canon reports that its Japanese employees [are] less likely to walk off with knowledge of the latest technological developments.[16]

More recently, the Japanese newspaper *Yomiuri* reported that Toshiba, Hitachi, Matsushita, NEC, and Renesas Technology, five large Japanese semiconductor companies, had reached an agreement to build a $1.6 billion joint fab for next-generation microprocessors. The deal was aimed at recapturing market share from US and Korean competition.[17]

In the US, Texas Instruments (TI), also a major semiconductor producer, announced in 2003 that it would build its next fab in Richardson, Texas. The company broke ground on the $3 billion facility in November, 2005 to produce digital signal processing (DSP) and analog-based system-on-chip (SoC) devices for wireless, broadband, and digital consumer applications in 65 nm process technologies on 300 millimeter wafers.

Rich Templeton, TI's president and CEO, explained the decision by saying, "Texas Instruments is one of the few semiconductor companies able to leverage the significant advantages that come with closely linking our chip design activity with manufacturing and process technology development."[18]

---

[16] I. Rowley with H. Tashiro, "So much for hollowing out: Japan's giants are investing in plants at home again. Why the switch?" *BusinessWeek online* (October 11, 2004), www.businessweek.com (accessed on February 10, 2006).

[17] Reported in C. Preimesberger, "Top Japanese chip makers reportedly ally to build plant," *eWeek* (November 18, 2005), www.eweek.com (accessed on February 10, 2006).

[18] From a corporate press release, "Texas Instruments prepares for future growth, sets November groundbreaking date for new facility; new fab to compete

Another indication of a commitment to chip manufacturing in the US is the plan of AMD to build a $3.5 billion plant in upstate New York. State incentives helped in financing.[19]

Clearly there is a future for electronic manufacturing in developed nations, provided it's the right kind of manufacturing.

## Innovation in manufacturing: A long-term project

We know that automation reduces the production costs for established product categories. The mathematical modeling is easy. But investing in automation is a more difficult decision in new markets. The payback is not as clear with manufacturing technologies that enable new products for emerging industries. This is where visionary corporate management comes in. Companies that invest in these processes must be prepared to wait a long time for any returns, in some cases more than a decade.

In most cases they are alone in pursuing the new technologies. Once the technologies mature and automated production equipment becomes commercially available, however, the pioneers can expect to see competitors emerge very quickly.

### *LCDs: Pioneers and profiteers*

Perhaps the best historical example of this situation is the flat-panel liquid-crystal display (LCD) industry. It took over twenty years for the technology behind this revolutionary innovation to mature, and a total of forty years for it to end up in mass-produced television sets.

When the early technology was developed at RCA Laboratories in the 1960s, it was only suited to small numeric displays, such as electronic watches. In fact, RCA set up a factory for the commercial production of small LCD displays for just this purpose. Everyone dreamed about the LCD's potential to replace picture tubes in television sets, but there were enormous technological hurdles to clear before this could be achieved.

Inventing solutions to these problems ultimately required the investment of many billions of dollars. None of the dominant

globally with 300-mm wafers, advanced processes," www.ti.com/corp/docs/press/company/2004/c04050.shtml (accessed February 17, 2006).
[19] Computer Wire Computergram, June 26, 2006.

manufacturers of television tube displays in the US (including RCA) were willing to commit themselves to such a program.

But one pioneering company, the Sharp Corporation of Japan, finally took the lead. Sharp made the investment in automated plants and engineering necessary to master large-scale LCD production technology. Much of this equipment had to be developed because no one had ever made such products before.

Their gamble paid off. In the 1980s new products appeared, unforeseen in the 1960s, that used LCDs to display information and graphics. The portable personal computer, the cellular handset, digital cameras, and a host of other applications requiring readouts turned LCDs into a major market. Sharp emerged as the world's leading supplier of LCD displays.

But it was not alone, and here we have the classic problem for pioneers. Other companies recognized the emerging market opportunity in LCDs. Production equipment became commercially available to enable building plants with lower costs. Sharp found itself in competition with new entrants into the market.

By the 1990s, when demand for PCs and cellular handsets was exploding, the technology risks had been largely removed, so the supplier with the lowest LCD production costs was likely to win the highest market share.

Companies in Taiwan and Korea began making massive investments in automated LCD manufacturing plants. With their focus on cost reduction, they soon overtook the Japanese vendors, and emerged as the leaders in the LCD market.

This was no small achievement: the display industry reached $60 billion in sales in 2004.[20] Figure 6.6 shows the shift in large-panel LCD market share from Japan to Korea and Taiwan between 2000 and 2005, with projections to 2008.

We should stress that the migration of this technology from Japan was only possible because of the strong engineering skills that had been developed in Korea and Taiwan. The availability of automated production equipment leveraging ever more sophisticated digital technologies makes it possible to manufacture LCD displays more

[20] Market statistics in presentation by Dr. A. Bergh, Optoelectronics Industry Development Association (OIDA), Washington, DC (2005). Figure courtesy of Displaybank.

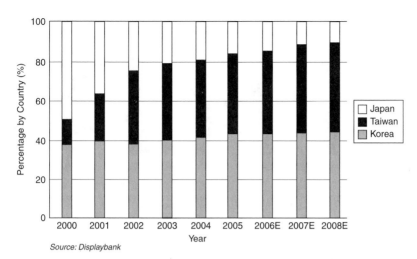

**6.6.** Large-panel liquid crystal display (LCD) production by country and by year. Reprinted with permission from Displaybank (ref. 20).

cost-effectively, but it is still a very difficult process, requiring a high level of technical knowledge.

To make the LCD used for a typical PC display, the manufacturer must deposit thin films on glass with over three million functioning thin film transistors (Chapter 1). Manufacturing these products is as challenging as making integrated circuits.

For the new market leaders, it was well worth the effort. Figure 6.7 shows the trend of production value of Taiwan's photonics industry. The market value of flat-panel displays (LCD, PDP, and OLED) and projectors has been growing steadily and comprises 30 percent of the global photonics market in 2004.[21]

The story is the same in Korea. In response to the growing market for large television displays, Korean companies such as Samsung have made large investments in their manufacturing facilities. In 2005 they emerged as major suppliers.

While these manufacturers are cost-competitive now, there is no guarantee they can sustain this position. In a world where technology

---

[21] Global photonics market and the photonics industry in Taiwan, www.pida.org. tw/newversion/homepage/2001new/english/overview.htm (accessed on June 12, 2006).

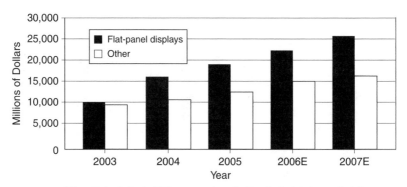

Other = Opto-electronic (OE) components, optical input/output devices, optical storage,
optical fiber communication (OFC), optical components, and laser applications

*Source: Photonics Industry and Technology Development Association (PIDA)*

**6.7.** The production value of Taiwan's photonics from 2003 (ref. 21).

diffuses rapidly and foreign competitors can offer lower labor costs and superior technology, keeping pace with automation is obviously important, but it is not enough to maintain a competitive advantage. Other factors come into play.

## Domestic specialties, offshore commodities

One of these is focusing on specialized products. Japanese companies have adopted this approach, translating proprietary skills into robust competitive positions while making optimum use of less expensive offshore production. Consider, for example, how automotive component suppliers balance the twin demands for technological sophistication and low-cost production.

Of course, the Japanese are famous for turning excellence in manufacturing into a major competitive advantage, but recent cost pressure from China has forced changes in this strategy, even for domestic production.

Toyota is supplied within Japan by a network of companies with interlocking ownerships. These suppliers manage to maintain domestic plants with highly-skilled workforces, even though their labor costs are about ten times those in China.

Their secret: unique products that can command premium prices. In Nagoya, where Toyota and its suppliers have major manufacturing facilities, the local plants continue to concentrate on high-end, difficult-to-make products. But the suppliers are shifting production of simpler components to China to give their big customer higher tech or lower labor costs, depending on which is more appropriate.

The Japanese machine tool industry follows a similar strategy. Easy-to-manufacture parts for the machines are made in low-cost countries. High-end parts continue to be made in Japan, with a constant focus on enhancements to maintain a performance advantage over overseas competitors. As much as 85 percent of a given product may use components from low-cost countries, but the proprietary 15 percent of the machine that creates the major part of its value is jealously guarded and produced domestically.[22]

As another example of how companies in capital-intensive industries site their manufacturing facilities according to the skill levels needed, consider a German company, Continental AG, a global supplier of tires and automotive components to Daimler-Chrysler and other automakers.

The company has moved labor-intensive operations offshore. However, it keeps its most automated operations in Germany and other countries with high labor costs, where it depends on small but specialized teams of skilled employees.

In Continental's German-based operations labor amounts to less than 4 percent of production costs. These plants work at full capacity and use extensive automation to produce high value-added products. Continental sees no advantage in moving such plants to low-wage countries unless logistical reasons require them to be closer to customers.[23]

## Proprietary processes

Corning, the world's leading supplier of specialty glass products, is a prime example of a company that has built its success on highly-specialized process automation. When large market applications for glass emerge, the company quickly invests in new production equipment and pioneers a dominant position in the new products.

---

[22] J. Sapsford, "Japan's economy gains steam from manufacturing heartland," *The Wall Street Journal* (October 11, 2005), 1.

[23] M. Landler, "A German auto supplier Delphi might envy," *The New York Times* (November 24, 2005), C1.

Corning spends 10 percent of its revenues on research and development, and relies on proprietary processes and a well-trained workforce to produce unique products. In the 1950s, Corning pioneered the production of TV picture tubes. In the late 1960s and 1970s, it led in the development of glass fibers for communications. In 2000, it introduced special glass plates for liquid crystal displays.

It has plants in Japan and Taiwan that produce the glass plates used to manufacture LCDs. Although these are both now relatively high-cost countries, the plants make good economic sense, because they are close to Corning's flat-panel display customers. Shipping large glass plates over longer distances would be far too costly.

In keeping with Corning's philosophy, each plant is automated. Both can produce $1 billion dollars' worth of product annually with only about 250 employees.[24]

## Offshoring and outsourcing software development

Chapter 2 introduced the innovations in software engineering that transformed programming from a craft to something more like a production process. These technological innovations turned software development into a global industry, though it remains a very labor-intensive process.

As with any product, the overall design of a software program is the key to its success and value. In standard practice a small number of skilled software architects designs the "solution," but a much larger number of engineers codes and tests the finished software. About 40 percent of the development cost of most programs is in the testing.

The common practice is for companies to retain the core design teams in their home countries but outsource the coding and testing chores to lower-cost engineering groups in other countries. The idea is to keep the teams that do the fundamental innovation close to the center of the business. This is also a good way to protect intellectual property. For example, core software development for Apple products is done in the US.

Designing a new software or software-based product is one thing. Development and support is another. Although improvements in

---

[24] P. Marsh, "A careful giant behind the glass," *Financial Times* (September 30, 2005), 10.

training and tools continue to increase the productivity of software engineers, development activity is still a major cost center for many companies, and occupies an ever-growing portion of the technical staff.

To reduce its cost, companies in all types of businesses have opened software engineering facilities in India or outsourced the work to contract software companies there and elsewhere. Many of these companies are service businesses, such as banks, insurance companies, and investment firms that maintain large proprietary software programs and databases.

But it is not just software-dependent businesses that need to cut back on software development costs. Equipment companies incur substantial programming costs as well. During product development they spend more on software engineers than they do on engineers in other disciplines. In fact, an estimated 70 percent of the engineering cost needed to bring a new electronic equipment product to market is spent on software development.

Large companies such as General Electric opened software development facilities in India in the 1980s. They were followed by many others. Today practically all major US companies that rely on software have offshore facilities employing many thousands of engineers.

However, wages are rising in India at about 10 percent a year. The escalation is prompting companies to look at sites such as China and Eastern Europe for less expensive programming services. So far, the fact that English is the language of instruction in Indian colleges has helped facilities there retain this business.

Other companies were also interested in having access to India's low-cost software resources, but either could not open their own facilities there, or chose to outsource such a small fraction of their software development projects that creating an Indian operation did not make economic sense. In response to this market opportunity, many contract software companies were established in India.

The founders were frequently engineers who had returned from the US. Some of these companies, such as Infosys, are now publicly traded and are highly profitable.[25] In 2004, the total revenues of the software contract companies in India are estimated to be in excess of $17 billion, having grown at a 30 percent annual rate.

---

[25] For a lengthy discussion of the outsourcing industry see R. Hira and A. Hira, *Outsourcing America* (New York: AMACOM, 2005).

It is estimated that they employ about one million engineers, although this number is hard to verify. These companies take full responsibility for delivering software in accordance with the specifications of their customers. Their clients own the code they produce.

Contract software businesses have been established in other countries as well, notably in Eastern Europe. One country where contract software companies have been emerging is Ukraine, where engineering labor rates are below those in India (see Table 4.1).

Then there are international contract software companies such as Ness Technologies, which provides services from locations in India and Slovakia. Eastern Europe is attractive because of its proximity to major Western European companies, and because many Eastern Europeans know German or French. This makes communications easier than dealing with the English-speaking Indian facilities.

We can expect that other countries will enter the contract software market as they expand their engineering populations. However, India is currently unique in the size of its software engineering force.

Looking at the frenetic activity in the contract software universe, it is fair to ask how much of the cost of product development outsourcing actually saves. From personal experience and from conversations with outsourcing companies, I estimate at this writing that software can be produced in India and other low-engineering-cost countries for about half the cost of doing it in the US or Western Europe.

This figure takes into account the increased cost of management and communications required for such offshore organizations. However, the savings are neither instant nor automatic. It takes time to reach full efficiency because of training requirements, and actually realizing the savings presupposes close management of all projects.

## What happens to outsourced workers?

It is not the purpose of this book to address social issues or stump for changes in government labor policies. However, no discussion of outsourcing can ignore the fact that moving software development (or any other kind of production) offshore does more than cut costs. As recent headlines show, it obviously affects domestic workers.[26]

---

[26] R. Hira, "Impacts and trends of offshoring engineering tasks and jobs," *The Bridge* (Fall 2005).

US Department of Labor Statistics show that 35 percent of the software workers displaced between 2001 and 2003 were still unemployed in January 2004. Of the 65 percent who had found employment, only 43 percent earned as much as they had before being laid off. The unemployment rate of electrical engineers and computer scientists has increased from between 2 to 4 percent in the years from 1982 to 2000, to between 4 and 5 percent in the 2001–2004 time periods.[27]

What is happening is that an increasing level of skill in emerging technological areas is required to maintain high wage levels in a developed economy. It is well known that there are very large differences in the productivity of software engineers. For example, while average coders find it hard to get jobs that pay well, highly-skilled engineers in Internet-related software development are in high demand in the US.[28] Furthermore, as long as corporations continue to keep the highest software development skills in their home countries, the demand will be high for very talented engineers to fill these roles, and their compensation will reflect their value.

We would do well to remember that our prosperity, and the domestic employment it brings, has been the outgrowth of innovations and the new industries they create. Creating a climate for innovation in all phases of industrial development, including manufacturing investment, may be the best prescription of all for the conditions that industries in the developed economies face today.

## The future of manufacturing

With the contemporary focus on software and the Internet as sources of value creation, it's easy to forget that the manufacturing of hard goods continues to be an important creator of value and source of employment.

Unfortunately, at least in the West, the future of manufacturing in general does not look bright. If current trends continue, all but the most highly-specialized manufacturing will soon leave high-cost countries.

[27] Displaced Workers Survey (2004), US Bureau of Labor Statistics, Washington, DC.
[28] B. Prabhakar, C. R. Litecky, and K. Arnett, "IT skills in a tough job market," *Communications of the ACH* (October 2005), 91–94.

And even in this case the availability of low-cost capital and government subsidies will affect the location of new plant construction.

The extent of the migration is already staggering. Mass-market consumer products invented in the US, such as personal computers, color televisions, flat-panel displays, cellular handsets, and even the iPod, are manufactured in Asia.

Some of the components used in these products, such as the semiconductor devices, are made in the US. To underscore the extent of Asia's rise in manufacturing, it is estimated that about 45 percent of the semiconductor devices shipped in the world in 2005 ended up in Asian countries, where they are assembled into products. Chapter 8 addresses this situation in more depth.

The semiconductor industry in developed countries is threatened too. As fabs come on stream in China and elsewhere, local vendors will increasingly fill the demand for chips in the newly constructed assembly plants. This is the manufacturing "network effect," where component production tends to move to the final assembly region when conditions are ripe.

To summarize, other Asian nations are now following the path blazed by Japan, using their cost advantage in labor and engineering to gain entrance to electronics manufacturing. Then, aided by the skilled use of automation, they move up the value chain to manufacture products of increasing complexity. But they are doing it much faster, thanks to the very electronic technologies they are making.

A total loss of manufacturing industries to lower-cost countries is not a foregone conclusion, as our examples have shown. Moreover, even in industries outside of high-tech, production workers in the developed economies will agree to lower wages when faced with unemployment.

We see this trend in the US, where union wages in some contracts are going to a two-tiered model. As reported in *The New York Times*,[29] Caterpillar, a major maker of heavy construction equipment company, will be paying new workers $12–13 per hour, whereas previously employed workers earned about $20 per hour more. This is a wage rate that the company believes allows it to compete while keeping production in the US.

---

[29] L. Uchitelle, "Two tiers slipping into one," *The New York Times* (February 26, 2006), 1.

But this is a coping mechanism, not a solution. A better answer to the migration of jobs lies in how industry approaches manufacturing.

Developed nations that do not invest in sophisticated manufacturing are missing major opportunities for value creation, over and above what they can get through product or technology innovation. Both US and Asian enterprises have demonstrated that relentless attention to developing specialized manufacturing technologies is an effective strategy for staying competitive in the field. More companies should emulate these efforts.

If LCD makers in Taiwan, or Corning in the US, can capture and keep a leadership position in manufacturing high-value products, others can as well. But this cannot be done by clinging to the past and subsidizing uncompetitive industries.

# 7 | *Your government is here to help*

S TEEL, textiles, clothing, and consumer electronics: one after another, manufacturing industries in the US and other developed countries have buckled under the pressure of low-cost foreign competition. It's happened so often over the last half-century that it no longer generates much political heat.

That doesn't mean governments won't enact protectionist measures to shield their basic industries. In 2002, for example, the US imposed tariffs on imported steel. But tariffs are often little more than rearguard holding actions. The basic industries (always excepting agriculture and big employers like automobile manufacturing) no longer seem to be regarded as central to the economic future of their countries.

Let foreign competition threaten the growth engines of a developed nation's economy, however, and this attitude suddenly changes. The government will muster an arsenal of legal and economic counter-measures to thwart the challenge.

Most of these measures target the protection of intellectual property (IP), the foundation of electronics-related industries. They include tactics such as embargoes and policing. Unfortunately, these tactics are not much more effective than import tariffs.

Other approaches, such as patents, trade secrets, and copyrights, may have more staying power. We will consider each of them.

## From embargoes to patent protection

The promotion of industrial development has never been a gentleman's game.[1] Countries building their industrial bases always seek to acquire valuable technologies by fair means or foul, and attract investment capital with government subsidies. Once they achieve their aims, they

---

[1] For a perceptive overview, see P. Choate, *Hot property: The stealing of ideas in an age of globalization* (New York: Alfred A. Knopf, 2005).

ır energies toward the task of keeping foreign competitors from
.he same thing to them.

ı. .s is the cycle we've seen in the rapid industrialization of Germany
in the second half of the nineteenth century, that of Japan in the 1960s
and 1970s, and the emergence of China in the 1990s. German industry
took business away from the British, the Japanese from Western companies, and the Chinese from Western and other Asian competitors. In
each case the incumbent leaders called for government protection from
the foreign challengers.

While the debate over protectionist policies is centuries old, over the
last half-century it has changed in one significant aspect: the current
international focus on protecting IP. This relatively new strategy shows
how important high-tech industries have become in developed economies. One source estimates that 42 percent of the $5 trillion GDP
generated in the US each year takes the form of intellectual property. IP
in this sense includes software, music recordings, and video products.[2]

All of these products are easy prey for software pirates, especially in
the consumer market. The stakes are huge. Although precise figures are
difficult to come by, electronics-based industries are clearly at risk of
losing IP that costs them billions of dollars to develop.

Governments use one or more of the following techniques to try to
stem the theft of IP and control where the technology goes:
- export embargoes and similar constraints: blocking certain countries
  or individuals from buying or accessing technology, usually for
  security reasons;
- legal protections for IP, including patents and copyrights;
- tariffs and other trade restrictions to protect local industry while it
  struggles to establish itself or become competitive again;
- prohibitions on importing products that are deemed (usually by
  court action) to violate patents or copyrights.

The first tactic, embargoes and access restrictions, applies mainly to
technology that has implications for national defense. This relates only
tangentially to our focus on innovation and economic growth, but it's
worth a brief look to see how government action works.

In an embargo, the US government (for example) might restrict the
sale of a technology to certain countries (such as Iran) because

[2] V. Shannon, "Value of US ideas is huge, but protection is lax, study finds,"
*International Herald Tribune* (October 14, 2005), 17.

permitting those countries to acquire the technology would create a security risk. The embargoed technology usually has important military or intelligence applications. The country imposing the embargo fears that the technology will find its way to adversaries, who could use it to compromise the defenses of the originating country and its allies.

Using this same rationale the US sometimes restricts foreign students and visitors from having access to such advanced technologies while they are working here.

Unfortunately, the effectiveness of such restrictions is questionable at best. For one thing, few technologies are totally domestic. Most advanced technologies have spread around the globe, making even embargoed products readily available outside the US. The only predictable effect of embargoes, therefore, is to steer potential buyers of embargoed technology away from US firms and toward their foreign competitors.[3]

In addition, restricting the technical activities of foreign students simply drives those applicants to other countries, which are just as anxious to attract overseas talent. This is not a desirable situation for the US. As pointed out in Chapter 4, immigrants are very important to US high-tech industries.

In short, government action to embargo IP often fails to achieve its goal, and can handicap the country's own industries.

## Hiding behind trade barriers

Governments will also try to protect their industrial base by creating trade barriers. They often do so at the request of an affected company or industry.

Trade barriers give companies or industries a temporary reprieve from economic pressures, but that often amounts to little more than postponing an inevitable collapse. Protection in and of itself cannot make a company more competitive.

### RCA and Zenith vs. Japan Inc.

That certainly proved true in the case of a famous American TV set manufacturer when Japanese competition invaded its home market.

---

[3] A. Segal, "Blocking the flow of ideas will not bring security," *Financial Times* (September 1, 2005), 11.

The drama played out during the 1970s, when Japanese companies entered the US market in force. They included such competitors as Sharp, Sony, and Matsushita (maker of Panasonic products, and in 1974 the buyer of Motorola's American-made Quasar TV business), and they arrived with a formidable array of low-priced, high-quality products.

RCA and Zenith, the two leading domestic makers, reacted very differently to the challenge. RCA, an industry pioneer, had a formidable patent portfolio going back to the 1920s in commercial radio and television technology. Continuing investments in research and development had kept the company at the forefront of both technologies.

Zenith, on the other hand, got its start in 1927 with a license to RCA's radio technology. Over the years it had invested very little in innovation. When consumer television emerged in the 1950s as a major new market, Zenith continued to rely on acquired technology. During the succeeding two decades, the company's patent wars with RCA and other competitors, combined with its failure to invest much in its own R&D, had left Zenith in a very weak technology position.

At the time of the Japanese onslaught, RCA's strategy was to license its technology globally. The resulting large annual licensing revenues helped finance more research and development, which then allowed RCA to continue developing highly competitive products.

Lacking a technology edge, Zenith took the path of seeking legal protection against the imports. In 1974 the company filed a federal suit charging Japanese television manufacturers with dumping and illegal collusion. Specifically, Zenith claimed that a Japanese television set selling for $500 in Japan, where the market was closed to imports, was being dumped for $300 in the US.

This highly visible and politically charged case wandered through the legal system for over a decade, at great cost to Zenith. Eventually, in 1985, the Supreme Court ruled in favor of the Japanese.

Zenith's effort to protect itself through trade sanctions had been in vain. Lacking competitive products, the company never managed to make a profit. It was eventually acquired by its Asian manufacturing partner, LG Electronics of South Korea, and disappeared as a separate entity.

In one final irony, Zenith did finally create its own technological legacy – just about the time it was being acquired. The company's engineers developed the 8VSB modulation scheme for terrestrial broadcasting of digital TV signals. This innovation has been adopted by the US and other countries that use the ATSC system for digital and high definition television.

## Protectionism Asian style

Charges of dumping and similar unfair trade practices are also prevalent in other segments of the electronics industry. Semiconductors are a case in point.

In the 1980s the Japanese chip industry entered on a period of aggressive expansion, evidenced by its growing exports to the US. This produced an outcry from American makers about Japan's dumping of dynamic random access memory (DRAM) chips in the US market.

Perhaps predictably, the plaintiffs got the same satisfaction from the government as Zenith did: the flood of low-priced, high-quality imported devices did not abate. Production of DRAM chips wound up being concentrated in Japan. Today Micron Technologies is the only US-based manufacturer still in the DRAM business.

But the story does not end there. Competitive battles between national champions are still going on. In the 1990s, Samsung of South Korea emerged as a worldwide leader in memory chips and selected areas of consumer electronics. Its ascendancy, and that of other Korean manufacturers, came primarily at the expense of the established Japanese companies.

A similar scenario is now playing out between China and Taiwan. Taiwanese semiconductor companies are producing non-memory chips in increasing volumes, and they are asking for government help to keep from losing domestic business to cutthroat competition from the mainland.

In August 2005, J. S. Huang, the director general of the Taiwanese Semiconductor Industry Association (TSIA), noted that Taiwanese equipment manufacturers imported 16.8 per cent of their chips from the mainland during the first half of that year. In terms of volume, China had also become the second-largest source of chips to Japan. Even more worrisome to the TSIA, these were high-performance chips, marking a break from the pattern of previous years when mainland imports were mostly low-end products.

Huang, obviously asking for government help, said that "Taiwan should prevent mainland Chinese dumping of chips on the island."[4] Given other political considerations between Taiwan and China, and the fact that lower-cost chips help Taiwanese equipment manufacturers stay competitive, it is unlikely that anything will be done in response to this plea.

## Nowhere to hide, somewhere to go

It has always been difficult to shield high-tech industries (or indeed any industry) from foreign competitors. Asking the government for import protection on the basis of unfair pricing tactics rarely accomplishes much in the US. It certainly does not provide a bulwark against a determined, well-capitalized competitor, especially when the challenger combines efficient manufacturing with state-of-the-art technology.

Trade barriers may delay the entry of foreigners, but in the final analysis, the only approach that creates sustainable businesses is to combine technological innovation with market savvy and creative ways to cut costs. That's the lesson we learned in Chapter 6 from the examples of the Taiwanese LCD makers, the IP-based US semiconductor companies, and the Japanese machine tool industry.

Another way to stay in the game is to form a joint venture with your potential competition. As Chapter 4 briefly mentioned, Alcatel, a French telecommunications company, has done just that. It has entered into a partnership agreement with a Chinese government-owned entity. Let us take a closer look at Alcatel's calculated move.

China is industrializing faster than any other country in history. Although hardly a paragon of laissez-faire capitalism, it attracts enormous amounts of outside investment capital. The Chinese government's industrial policy targets the development of specific industries, with information and electronic technologies getting high priority.

All of this spells trouble for companies in those industries in other parts of the world. As Chinese manufacturers get better at making world-class products and begin to export, established companies like Alcatel are faced with an age-old dilemma: how do you survive when

---

[4] "Organization warns of likely mainland Chinese chip dumping in Taiwan," *The Taiwan Economic News–China Economic News Service* (August 29, 2005).

your overseas competition can undercut your prices even in your home market?

For most companies, regardless of their nationality, their short-term answer has been to join the party. Don't rely on import barriers. Instead, move your most labor-intensive manufacturing operations to lower-cost countries. Keep the others close to home where the intellectual property can be better protected.

In many cases, companies abandon local manufacturing altogether. In the long term, as we discussed in Chapter 6, intellectual property generated by manufacturing excellence creates value in many industries, but companies under financial pressure to reduce costs fast may not be able to justify investing toward that objective when the option of moving offshore is open.

That's the reason behind the rapidly growing influx of foreign-financed production plants into China. They are capitalizing on the labor cost savings available in emerging economies, the better to compete on international markets.

There is an alternative to building those plants, however, and the Chinese government is heavily promoting it – a joint venture. This arrangement allows international equipment companies to leverage advanced technology with low-cost manufacturing, so they can produce exported products at costs far lower than otherwise possible. This is the route chosen by Alcatel.

Alcatel Shanghai Bell (ASB), a joint venture formed in 2001, is 49 percent owned by Alcatel. A Chinese government-owned entity controls the rest. ASB produces telecommunications network equipment in China using Alcatel technology.

The arrangement gives ASB access to China's low-cost resources (labor and infrastructure). It also qualifies ASB for subsidies in the form of R&D incentives and low-interest loans to finance exports. Essentially the joint venture gets the same preferential treatment as other similar domestic companies, including access to assistance from municipalities in industrializing regions to encourage local employment.[5]

With these advantages ASB is well positioned to sell into the fast-growing local market for communications equipment, and move into international markets as well. ASB gives Alcatel the ability to compete

---

[5] This information is from a speech by Olivia Qiu, Vice President ASB, July 2005.

on price and product quality anywhere in the world, with Alcatel providing the international marketing reach.

But Alcatel and other vendors face a difficult choice – either take advantage of the Chinese partnership opportunity and compete world-wide, or fight a losing battle for market share with emerging Chinese companies such as Huawei, which offer state-of-the-art products built with the same low labor costs and access to financial resources that benefit ASB.

While this form of government intervention has its advocates and its advantages, it hasn't been around long enough to build a track record. Only one type of government program has established a positive record for developing sustainable industry: the patent system. It also has the advantage of being built around the concept of encouraging the development of innovation, in the form of IP.

## Protecting IP: The new international battlefield

The patent system in the US was created over 200 years ago. Its intent was to promote industrial innovation by giving a limited monopoly to inventors. Few people debate the historical value of the system in fostering R&D investments, but recent changes in the patent granting process have prompted a great deal of criticism. Many experts charge that the patent system does as much to hinder as to promote innovation.

By law, a patent is only granted for inventions that are useful, new, and not obvious. The definition of "obvious," however, is a major bone of contention in most patent suits. The intent of the law in excluding "obvious" innovations is to avoid patenting trivial extensions of the existing art. This, of course, begs the question of what is trivial, and opens the door to more challenges over the legitimacy of patents.

The courts have been heavily involved in settling such disputes almost from the beginning of the system. Resolving this issue is a process that enriches lawyers while it clogs the courts. The demand for adjudication was so great that in 1982 a special court was created to handle certain types of patent case appeals.

According to many experts, however, the biggest problem with the current patent system is not the number of disputes, lawsuits, and appeals it generates. The real difficulty is that patents have become too easy to obtain, especially since the 1980s, when the criteria for awarding them were greatly expanded.

It's hard to determine the exact dimensions of this problem. Ideally we would like to compare how patent approval rates have changed over the years. But it's difficult to calculate an accurate figure for even one year, in large part because inventors can file multiple patent applications for the same work.

However, we can compare approval rates in the US to those of other countries. Between 1994 and 1998, according to the US Patent and Trademark Office (USPTO), about 75 percent of original patent applications in the US were eventually granted. The rate of acceptance in Japan and by the European Patent Office is substantially lower – in the 50 to 60 percent range.[6] The high acceptance rate in the US has serious implications for the future of innovation.

## Moving software into the patent system

The largest single change in the US patent system in the 1980s, and the one with the biggest impact, was allowing patents for software. Up to this point software IP had been protected by copyrights on a program, or by keeping the code a trade secret.

The landmark case that opened the way to patenting processes implemented in software was *Diamond* v. *Diehr and Lutton*. This dispute (450 US 175, 209 USPQ 1) went all the way to the US Supreme Court in 1981.[7]

Diehr and Lutton worked for a company that made products molded from rubber compounds. The standard method for molding these products was to place the compound on a heated press, set a timer to sound an alarm at the end of the correct curing period, and have an operator remove the finished product from the press.

---

[6] Information on the US patent system was drawn from S. S. Merrill, R. C. Levin, and M. B. Myers (eds.), *A patent system for the 21st century* (Washington, DC: The National Academies Press, 2004), p. 53; and from A. B. Jaffe, "The US patent system in transition: Policy innovation and the innovation process," *Research Policy* 29 (2000), 531–537.

[7] J. Tripoli, "Legal protection of computer programs," *RCA Engineer* 29 (January/February 1984), 40–46; and R. Hunt and J. Bessen, "The software patent experiment," *Business Review* Q3 (Philadelphia, PA: Federal Reserve Bank of Philadelphia, 2004), 22–32.

Diehr and Lutton's contribution was programming a computer to automate the process. They developed a program with the time/temperature relationships for curing specific molding compounds, which was fed into the computer during the curing process. Calculations of data on temperature and elapsed time using a stored algorithm allowed the computer to make continuous adjustments during the curing process. At the end of the calculated correct curing time the computer triggered a switch that opened the press.

Their patent application was initially rejected by the USPTO on the grounds that patent protection did not apply to computations using mathematical algorithms. Algorithms were public domain. In its landmark decision the Supreme Court allowed the patent, based on a ruling that the application involved not just a calculation using an algorithm, but associated machinery operations for controlling the cure process, such as temperature monitoring and opening the press.

In other words, it was not the computation, but its use in managing a physical process that was the deciding factor in allowing the patent. Effectively the court was saying that one had to look at the whole patent claim, not just the mathematical algorithm.

The criteria for patents on processes implemented in computer software (usually referred to as software patents) were further clarified by subsequent court cases. These cases also eliminated prohibitions against patents on business methods and against patents that included a mathematical algorithm that had been part of earlier judgments.

To an engineer this meant that if you could draw a flow chart for your new process, you had something patentable. This was provided, of course, that your invention was novel and not obvious. The effect has been an explosion of software patent filings, with potentially chilling effects on innovation, as we will discuss below.

In retrospect, the problem was not that the USPTO was issuing patents on processes valuable to commerce. The real issue was that the quality of the patents was so poor.

The reason for this was the patent examiners' unfamiliarity with the field of software. They did not always understand the prior art, which they must review to determine whether a patent claim is both novel and not obvious. The USPTO has recognized this shortcoming and is addressing it.

## Patents, copyrights, and trade secrets

After 1990 the floodgates opened wide. As shown in Figure 7.1, the number of issued software patents, which had been increasing slowly up to that point, jumped from 5,000 in that year to 25,000 in 2002.[8] Software patents accounted for 15 percent of all patents issued in 2002, up from less than 6 per cent in 1990 (see Figure 7.2).[9]

Patents provide new and potent protection for software IP. Under the law, courts can issue injunctions against any company found guilty of patent infringement. In effect, the infringing business can be shut down if the patent holder refuses to license the use of the contested patent or make some other arrangement with the infringer.

Even large companies with battalions of lawyers are vulnerable to such penalties. At the end of a celebrated dispute, Kodak was forced to shut down its instant photography business when it was found to infringe a number of Polaroid patents.

To make matters worse, such cases may take years to decide, depressing an ongoing business under a cloud of uncertainty. A recent US Supreme Court ruling that judges have flexibility in deciding whether to issue court orders barring continued use of a technology after juries find a patent violation may result in much fewer injunctions – a welcome development for industrial activity.[10]

Patents are one of three mechanisms available to protect software. The others are copyrights and trade secrets. Table 7.1 compares how each works to protect IP.[11]

We think of copyrights as applying mostly to books, images, movies, and similar works. Prior to the *Diehr and Lutton* case, however, copyright was the most common method of protecting software, as covered under US federal law (Title 17 of the US code).

The ability to copyright computer code was established by Congress in the Computer Software Copyright Act of 1980. This law states that "a set of statements or instructions to be used directly or indirectly in

---

[8] Hunt and Bessen, "Software potent experiment," 24.     [9] *Ibid.*, 25.

[10] J. Bravia, M. Mangalindan, and D. Clark, "eBay ruling changes dynamic in patent-infringement cases," *The Wall Street Journal* (May 16, 2006), B1.

[11] J. Tripoli, "Legal protection of computer programs," *RCA Engineer* 29 (January/February 1984), 45.

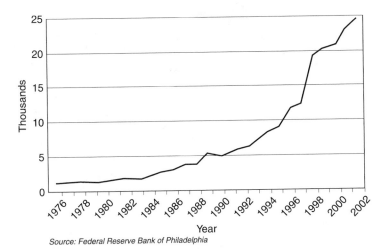

Source: Federal Reserve Bank of Philadelphia

**7.1.** Software patents granted in the US by grant date (1976–2002). The data comes from the USPTO as computed by Hunt and Bessen, "Software patent experiment." Reprinted with permission from the Federal Reserve Bank of Philadelphia (ref. 8).

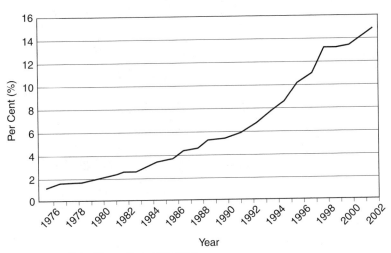

Source: Federal Reserve Bank of Philadelphia

**7.2.** The percentage of patents granted in a year that are software patents (1976–2002). The data comes from the USPTO as computed by Hunt and Bessen, "Software patent experiment." Reprinted with permission from the Federal Reserve Bank of Philadelphia (ref. 9).

*Table 7.1 Brief comparison of software-protection techniques in the US*

| Parameter | Patents | Copyrights | Trade secrets |
|---|---|---|---|
| What is protected? | Underlying ideas, concepts, or methods used in the system. | Original works of ownership fixed in any tangible medium of expression. | Formulas, ideas, devices, compilations, programs, techniques. |
| How is protection obtained? | File an application in US Patent Trademark Office. | For published work: Place copyright notice on work. For unpublished work: Automatic upon completion. | Maintain secrecy. |
| When does protection start? | Issuance of patent. | Upon completion of the work. | At creation of secret information. |
| How long does protection last? | 20 years from date of patent filing. | Life of author plus 50 years. Work for hire: 75 years from publication, or, 100 years from creation, whichever comes first. | As long as information is kept secret. |
| Criteria for providing protection. | Must fall within statutory subject matter and must be new, useful, and unobvious. | Any original work of authorship. | Must have value and not be generally known to others. |

*Table 7.1 (cont.)*

| | | | |
|---|---|---|---|
| Who infringes? | Anyone who, without authorization, makes, uses, or sells claimed invention. | Anyone who, without authorization, copies, makes derivative works, publicly performs, or distributes copyrighted work by sale, lease, rental (subject to statutory exceptions). | Anyone who violates a confidential relationship or uses unlawful means to gain access to trade secret information. |
| Available remedies for infringement. | Injunction, reasonable royalties. | Injunction, statutory damages, royalties. | Injunction, damages, criminal sanctions for theft. |
| Applicable law. | Federal Patent Act. | Federal Copyright Act. | State trade secret statutes. |

*Source:* From Tripoli, "Legal protection of computer programs," ref. 11.

a computer in order to bring about a certain result" can be copyrighted. Hence, the program developed by Diehr and Lutton could have been protected by copyright.

As Table 7.1 shows, establishing a copyright on one's work is simple. The work is copyrighted as soon as it is "fixed in any tangible medium." At that point the familiar © symbol gives the public clear notice of one's claim to copyright.

This informal copyright does not entitle the holder to legal redress against IP theft. In order to initiate action against an infringer, the copyright holder must officially register the work with the Registrar of Copyright.

Copyrights also provide more limited protection than patents. A copyright protects the code that makes up a program, but not the idea behind the software. If someone else achieves the same results using different code, this is not considered a copyright infringement.

A patent, on the other hand, protects the concept of the software. Any other program that achieves the same results as a patented program would be regarded as an infringement on the patent.

A third way of protecting work is through trade secrets. The Uniform Trade Secrets Act, a statute used by some states, defines a trade secret as information, including a formula, pattern, compilation, program, device, method, technique, or process that:

1. derives independent economic value, actual or potential, from not being generally known, and not being readily ascertainable by proper means to other persons who can obtain economic value from its disclosure or use; and
2. is the subject of efforts that are reasonable under the circumstances to maintain its secrecy.

In order to treat information as a trade secret, companies must take steps to restrict access to the information and tightly control its dissemination. In the case of computer software, this is typically implemented in a license accompanying the product, whether a written agreement signed by the licensee, a printed license on the packaging, or a click-to-accept license.

Patent protection is obviously the strongest of the three protection techniques, but filing for a patent has its downside. The patent must disclose valuable information about the invention.

This can allow competitors to achieve the same results by designing their way around the patent. For that reason manufacturing industries prefer trade secret protection, which allows them to avoid disclosing process information. Other industries, including software, increasingly file patents as a defensive measure.

## Broader patents, more litigation

Patent litigation is a costly proposition. Since millions of patents continue to be filed around the world, just about every technology-based business will face it sooner or later.

Usually it's sooner. Depending on a number of factors, the life of a US patent is currently limited to between seventeen and twenty years (twenty years being most common), and patent holders want to derive maximum value from their IP before their exclusivity expires. For that reason most will quickly take action against any suspected infringements.

There's a broader basis for suits, too. It is commonly argued that easing the patent award process in the US has resulted in filings that are indefensible on the basis of novelty or other meaningful criteria. Software patents in particular, if written broadly enough, can cover almost anything under the sun.

Patents that cover services rendered through the use of software, such as the delivery of a service with a special billing method, are particularly onerous, because they can be extremely broad and difficult to litigate. There are companies whose only business is enforcing their claims to just these kinds of patents. Their stock in trade is lengthy, costly litigation. This tactic can hinder the development of a new service sector for years.

Two basic issues come up in patent litigation: first, whether the claims of the patent are valid; and second, whether a valid claim has been infringed. During the hearing, expert witnesses are called upon to testify about the infringement.

Regarding the primary question, that of validity, there are many issues that can arise, particularly whether there is prior art which could invalidate the patent in question. Those challenging the patent can request reexamination of the patent by the USPTO, which may reopen the question of patentability. This results in one of three outcomes: the patent can be upheld, its claims may be modified, or it may be found invalid.

There have been cases in which the Commissioner of Patents has ordered the reexamination of certain software patents which simply used a computer to implement a preexisting process.

The law states that "patent[s] shall be presumed valid." This is a significant advantage for incumbents, because the tendency is for courts to rule in favor of patent holders. Still, because so much depends on the testimony of expert witnesses, and on the ability of lawyers to explain highly-complex technical matters to non-technical judges and jurors, patent litigation is always risky and its outcome highly uncertain. More often than not the two sides reach a settlement just to avoid the hazards of jury trial.

Under this system new enterprises are obviously at risk from incumbents determined to "expense" them out of business through patent litigation, frivolous or not. So the question is: does the current process, which makes patents on new technologies easier to get, promote or hinder innovation and industrial growth?

There is no simple answer, and scholarly work on the subject is inconclusive.[12] One thing is certain: IP litigation is a growth business all on its own.

## How litigation can stifle innovation

It is understandable that companies that invest in research and development are anxious to protect their innovations by aggressive legal means. In a world where ideas move so quickly, this is a fair way to gain an advantage.

However, allowing trivial patents in a legal system that encourages litigation will have a chilling effect on innovation. The current system allows almost any innovation to be challenged if the challenger is willing to spend money on costly litigation.

Patent suits thus become a restraint of trade, delaying the introduction of new products or services while cases drag through the courts. This makes intellectual property litigation one of the major risks facing entrepreneurs (and investors).

The system also favors companies with the resources to aggressively fight patent cases. High-priced law firms often offer executive seminars

---

[12] Pertinent documents on the subject include J. Bessen and R. M. Hunt, "An empirical look at software patents," Working Paper No. 03-17/R, Federal Reserve Bank of Philadelphia meeting, March 2004; and R. J. Mann, "The myth of the software thicket: An empirical investigation of the relationship between intellectual property and innovation in software firms," American Law and Economics Association Annual Meeting, May 7–8, 2004, Paper 44.

on litigation as a "business strategy," where they tell big, entrenched companies that suits filed against potential competitors are a relatively inexpensive way to hold onto their markets.

It's not just big companies defending large market shares that benefit. As noted above, the system also encourages the formation of companies set up solely to buy broadly-based patents, which they use in litigation and nuisance suits in the hope of being paid to go away.

A well-publicized case, settled in 2006, involved Research in Motion (RIM), the company that provides wireless e-mail service to about three million BlackBerry handset subscribers worldwide. RIM was sued for patent infringement by NTP Inc., a company that owns certain software patents and whose sole business is deriving licensing income from them.

When the US Court of Appeals in Washington DC rejected the appeal of RIM on October 7, 2005, the company faced an injunction shutting down its operations unless it reached an agreement with NTP. At this stage of a case, with all appeals exhausted, the patent holder has all the cards.

Although RIM said it had developed an operating system that avoided the NTP patents, the OS was not ready for rollout, and RIM could not afford to risk its business or its customers. It had no choice but to settle – for $612 million. Technology columnist Michael Miller reacted to the settlement this way:

And the US Patent system continues to look pretty silly, as the Patent Office has overturned a couple of NTP's patents, but not until after the court case (and others are still being appealed). I've yet to be convinced that software patents make sense, given the vagueness of the difference between an idea and execution of that idea in the software world; and the lack of a good database [of] prior art.[13]

Unfortunately, the RIM/NTP case is just one of many such actions flooding the courts. And this is a long-term problem, not a temporary aberration. Two statistics prove it.

Figure 7.3 shows that the number of district court patent cases has more than doubled between 1988 and 2002.[14] As software cases

[13] "Patently crazy – The BlackBerry settlement," http://blog.pcmag.com/blogs/miller/archive/2006/03/06/682.aspx (accessed March 8, 2006).

[14] Merrill, Levin, and Myers (eds.), *A patent system*, p. 32.

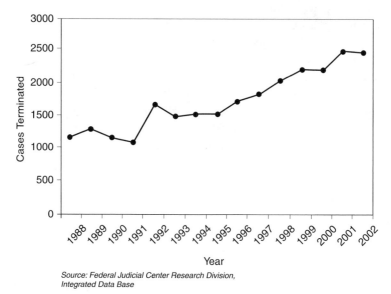

Source: Federal Judicial Center Research Division,
Integrated Data Base

7.3. Federal district court patent lawsuits terminated by fiscal year. "Terminated" includes judgments, dismissals, settlements, transfers, and remands. From Merrill, Levin, and Myers (eds.), *A patent system* (ref. 14).

proliferate, this curve is only going up. The increased legal activity has produced great employment opportunities for lawyers specializing in IP matters. Between 1984 and 2001, as Figure 7.4 indicates, membership of the American Bar Association grew by 29 percent while the membership devoted to Intellectual Property matters grew by 79 percent.[15]

Faced with these facts, what choice does an innovative business have but to play the game and file for as many patents as possible? At best, it is a means of keeping competitors at bay, and creating opportunities to cross-license patents. At worst, it can be a defensive tactic, establishing a claim that the company can offer to trade when attacked by others.

After years of ignoring the problem, even very big software companies are adopting this strategy. The biggest of them, Microsoft Corporation, until very recently relied mostly on market muscle and

---

[15] *Ibid.*, p. 33.

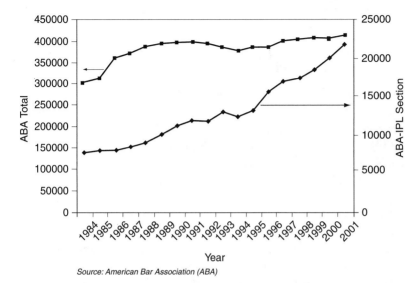

**7.4.** American Bar Association membership: Intellectual Property Law Section and total. Copyright © 2006 by the American Bar Association. Reprinted with permission (ref. 15).

copyrights for its competitive advantage. It has now decided to seek patents aggressively, with a target of 3,000 a year.[16]

Since Microsoft is the most dominant company in the software industry by far, this could spell big trouble for smaller innovators. If an industry includes a small number of large companies of equal weight, there is little incentive for extensive patent fights. The big adversaries can reach a mutually satisfactory agreement because both sides have something to trade and something to gain.

Things are different when a single company's IP controls a new industry, the way RCA's patents did in the early years of color television, or Qualcomm's CDMA technology does in the wireless world today. In this case a patent portfolio becomes the weapon of choice for freezing out competitors or extracting hefty license fees. Newcomers

---

[16] R. Stross, "Why Bill Gates wants 3,000 new patents," *The New York Times* (July 31, 2005), BU3.

must invent their way around the established patents and be willing and able to defend themselves in multi-million dollar litigations.

## Reforming the system

There is no shortage of people willing to complain about the US patent system. Pundits in every industry loudly proclaim that the system is "broken," and that patents are impeding innovation rather than promoting beneficial research and development.[17] But changing the system is a lengthy process. It is also likely to be very contentious. This is not surprising, since any reform will create winners and losers.

Nevertheless, there are signs of movement. In 2005 a bill was proposed in Congress to change the process of issuing patents in a way that would substantially reduce the differences between US and foreign practice. It would also subject patent applications to more careful scrutiny.

The hope is that the proposed reforms will halt the growth of patent litigation. If the bill is passed, its most far-reaching provisions would:

1. Start the time clock for establishing priority of invention from first to file rather than first to invent.
2. Eliminate the requirement that the patent disclose the best mode of practicing the invention.
3. Provide for publishing all applications.
4. Provide for post-issuance opposition procedures.

Not everyone agrees that the system needs reform. A minority view, represented by Bruce A. Lehman, former US Commissioner of Patents and Trademarks, holds that the current software patent policy in the US actually encourages innovation.[18] The more common view, however, and one with which I strongly agree, is that present practice encourages litigation, restrains competition by penalizing new companies, and diverts money to legal maneuvering that would be better spent on product development.

---

[17] See the summary in P. Samuelson, "Why reform the US patent system?" *Communications of the ACM* (June 2004), 19–23. Also informative is the work of A. B. Jaffe and J. Lerner, *Innovation and its discontents: How our broken patent system is endangering innovation and progress and what to do about it* (Princeton: Princeton University Press, 2003).

[18] B. A. Lehman, "Don't fear software patents," *The Wall Street Journal* (August 30, 2005), B2.

There is no question that the best and simplest way of dealing with frivolous patents is to ensure more careful screening by patent examiners. The examiners must do a more thorough job of checking prior art and testing the relative merits of the applications. Unfortunately, this entails more expense at the USPTO.

Would a stricter process for issuing patents, incorporating the changes discussed above, cut down on the amount or cost of patent litigation? Probably. It would certainly reduce the number of trivial patents granted, and also remove some of the more subjective issues in patent litigation.

To keep things in their proper perspective, let us recognize that the great majority of patents are never contested because they are essentially useless. Litigation actually involves a very small number of issued patents. This makes me wonder whether a tighter screen at the original issuing process would produce a meaningful reduction in the legal fights that surround the few important cases. But at least it would take some of the more frivolous "nuisance" lawsuits off the docket.

## An issue that has no borders

We've been focusing on patent issues as they relate to the US, but other countries are facing problems of their own. Patent standards differ from country to country, creating fertile ground for continuing legal conflicts. In those countries, as in the US, the growing economic importance of software guarantees that software patents will be at the heart of many disputes.

These countries are not much farther along than the US in addressing patent issues. An attempt to produce a uniform European Union software patent policy, one that would override national differences, was recently the subject of a three-year legislative battle within the European Parliament.

The proposed law would have granted patent protection to new, non-obvious, and technically meritorious software inventions. The arguments pitted open-source software advocates, who denied the need for software patents, against companies with large portfolios who were fearful of losing their future income from software IP.

The law was defeated in the Parliament in 2005. Hence there are no uniform standards within EU countries, and certainly no match between the patent criteria of the individual countries and those of the US.

Software will continue to be patented in the US and in EU countries, but standards of acceptance will vary. This will undoubtedly leave lots of room for litigation and uncertainty regarding what is protected, as one case after another makes its dreary way through the courts.[19]

## Patents and industry standards: A difficult marriage

One way that companies share IP is through industry standards. These are essentially technology platforms that ensure that products and services that comply with the standards fit their purpose and perform as intended. Examples include the various Motion Picture Experts Group (MPEG) software standards for encoding and decoding digital video, and the WiFi standards for wireless communications.

A standard normally includes IP from a number of companies. Each contributor is paid a royalty when other companies build products around the standard. The IP makes it easier to commercialize a technology by assuring interoperability and enabling faster design. Standards users don't have to re-invent the core technology, because it's already in the standards.

Since standards can involve hundreds of industry participants, a crucial question is how to deal with the patented IP incorporated into each standard. Standards-setting bodies address this issue head-on, requiring a participant to disclose whether a proposed contribution is covered by patents. This eliminates surprise claims for royalties from an individual company after the standard has been set. To be included in an IEEE standard, the contributing company makes a statement that the IP should be licensed under reasonable terms.[20]

Although industry standards incorporate agreements on their content from all participants, there are still squabbles among the competing companies. An industry thought leader with a strong patent portfolio that commands hefty license fees under the standard is a tempting target for complaints of unfair use of patent protection.

Take the case of Qualcomm, a company with a large number of core patents covering WCDMA technology for wireless communications. In a 2005 complaint before the European Commission, six major players

---

[19] "A sorry software saga," *Financial Times* (July 7, 2005), 12.
[20] C. K. Lindell, "Standard issue," *TechConfidential* (May/June 2005), 41.

in wireless (Broadcom Corporation, LM Ericsson Telefon AB, NEC Corporation, Nokia Corporation, Panasonic Mobile Communications Co. Ltd., and Texas Instruments) jointly alleged that Qualcomm had not lived up to the commitment it had made during standards development. The companies said Qualcomm had failed to license its technology on "fair, reasonable and non-discriminatory terms" to anyone that wanted to make equipment that complied with that standard.

The prospect of royalties from IP in an industry standard can touch off litigation well before the process of defining the standard even starts. This is particularly true if the standard is going to be essential to the new technology finding its market. A good example of this syndrome is radio frequency identification (RFID) technology.

RFID uses tags on products which, when interrogated by radio frequency signals, send back information. Consumers are familiar with highway toll passes which use this principle, but the technology is also an attractive replacement for the familiar bar codes widely used to track products through the supply chain.

The major benefit is that light sensing, necessary for using bar codes, can be replaced by signals which will be detected even in products that are enclosed and inaccessible to light. In 2004, Wal-Mart Corporation indicated its intent to mandate that RFID tags be part of its supply chain management process, and announced that its suppliers would be expected to adopt it.

The idea behind this technology is over thirty years old, but implementations have been hindered by the cost of the tags and reliability issues in reading them. Given Wal-Mart's immense size and buying clout, however, the chain's announcement was expected to spur rapid improvements in the technology and its widespread adoption. The announcement set off a scramble for ownership of relevant intellectual property.

It is estimated that as of 2005, over 1,400 patents have been issued in the field of RFID. Since 1976, by comparison, only 3,300 applications have been filed. Industry standards are essential for the adoption of a technology like this, and those who hold patents that are incorporated into the standards can realize valuable revenues when the licenses are negotiated.

But because of RFID's long history and the many patents filed, sorting out IP ownership is no trivial matter. Companies are now using legal action to establish their claims in preparation for licensing

IP for use in forthcoming standards. The first litigation over this
technology involved UNOVA, Intermec, and Symbol Technologies.
They initiated suits and countersuits in 2004 and 2005 in the US
District Court of Delaware.

## Producers and pirates: The world of digital media

When voice and video were analog, and the equipment and media for
recording them were bulky, expensive, and localized, the theft of
music, movies, and TV shows was not a major concern. Software
piracy wasn't an issue either, because there wasn't any software.

How things have changed, thanks to digital technology. Piracy of
recorded entertainment, video games, and computer programs is pervasive. Inexpensive technology allows criminals and consumers alike to
capture, store, and transmit voice and video anywhere in the world.

For the media industries that produce this content, the digital age is both
positive and negative. The positive: digital technology promises to transform the industry by creating new distribution channels for their product.
The negative: the industry is losing control of the product and its price.

### *Legal inroads: Digital distribution models*

Take a look at how revolutionary new forms of distribution are under-
mining the status quo in recorded music. The advent of inexpensive
portable devices for storage and playback of digital music (such as the
Apple iPod®, which had 80 percent of the market in 2005) has made it
possible to circumvent the conventional routes of disk sales and radio
play and instead transmit tunes directly into the hands of the public.

The enormous popularity of these devices concentrates a big portion
of the music distribution function in a few new hands. The digital
moguls who control this new distribution channel can set the price
for downloading songs, and force copyright holders to agree to their
terms.

It can be argued that new channels for the music industry generate
new revenue opportunities, and expand the customer base.[21] The good

---

[21] J. Leeds, "Apple, digital music's angel, earns record industry's scorn," *The New York Times* (August 27, 2005), 1.

news for the music industry is the fact that the sales of musical downloads totaled $790 million in the first half of 2005, up from $200 million for the same period in 2004. This amounts to 6 percent of industry sales, according to the Federation of the Phonographic Industry.

During that same period in 2005, recorded music sales (CDs and tapes) fell 1.9 per cent, or about $200 million, according to *The Wall Street Journal*. So the net effect is that digital downloads have actually increased consumer access to music – a great example of how new technologies can expand entertainment markets.[22]

The bad news is that the industry has to share its profits with newcomers, and that its album-oriented marketing strategy has been subverted. The music industry had effectively killed the sales of individual songs (the "singles" of years past) and forced the public to buy multisong albums at a huge price premium. With online music the industry must once again accept sales of singles, at one-twentieth the cost of an album.

## Illegal distribution: From DVDs to downloads

Even worse news comes from rampant piracy, which is stealing potential revenues in ever-increasing amounts. Piracy has always been a problem, particularly in emerging economies, but never to the extent it is now. It is estimated that in Russia alone the piracy of motion pictures, music, software, and books cost US companies about $1.7 billion in lost revenues in 2004.[23]

Over the next few years the situation will get rapidly worse for the music, video game, and movie industries, as pirates use increasingly sophisticated Internet-based electronic distribution techniques. Consider, for example, the comparative inefficiency of today's business model for illegally copying and distributing movies, and then look at the potential of the Internet to improve the picture – for the pirates.

---

[22] "Digital music's surging growth offsets CD slump," *The Wall Street Journal* (October 4, 2005), A19. However, worldwide music industry sales have declined. See J. Leeds, "Music industry's sales post their sixth year of decline," *The New York Times* (April 1, 2006), C2.

[23] G. Chazan, "In Russia, politicians protect movie and music pirates," *The Wall Street Journal* (May 12, 2005), B1.

In either case the process starts with a recording. Pirates first capture movies off the screens in theaters using video camcorders. Under today's model they then transfer the movies to DVDs with replication technology that is readily available (just look at the ads for disk copiers in any hobbyist magazine). The resulting disks are then sold at a fraction of the price of the authorized version.

It's not easy to interdict this form of piracy, but it is possible. DVDs are physical products. Their sale requires shipping to the target market. If you find the bootleg copies, you can destroy them. Even if the pirates slip them past authorities, they can't afford to ship them very far, so damages are somewhat contained by geography.

Electronic distribution via the Internet changes the game completely. Since there is no need to mechanically reproduce a physical copy of the movie, distribution can be both unlimited and invisible. When a boot-leg movie is just one more bitstream within international network traffic, it is extremely difficult to track and control.

It's happening now, and the economic effects are already being felt. The US movie industry reported global revenues of $84 billion in 2004. In that same year, when Internet distribution was still in its infancy, the industry's loss due to online bootlegging was estimated at about $3 billion.[24]

Movies in theatrical release are not the only targets of piracy. The cable TV industry is under attack as well. Programs recorded right off cable feeds are being distributed over the Internet by international pirates using peer-to-peer software. There is no technological solution to prevent this kind of piracy, since it simply relays programs that are readily available on cable systems.[25]

Given that the number of Internet nodes in the world is approaching the billion level, and that growing broadband penetration is making it possible for larger numbers of people to download huge, movie-size files to their computers, the field is wide open for pirates to push contraband to the public. Clever distribution technologies like BitTorrent only make it easier.

---

[24] T. L. O'Brien, "King Kong vs. the pirates of the multiplex," *The New York Times* (August 28, 2005), Section 3, 1.
[25] G. A. Fowler and S. McBride, "Newest export from China: Pirated pay TV," *The Wall Street Journal* (September 2, 2005), B1.

There are huge economic and cultural consequences to the growth of content piracy. Unlimited free distribution of content could destroy the industry. The average production cost of a feature length movie is in excess of $50 million. If pirates make it impossible to recoup that investment by siphoning off legitimate revenues, investors would quickly conclude that it makes little economic sense to put money into making movies.

The creative side of the business would likewise find it hard to justify investing time and talent into projects that brought them little or no financial return.

## Government and industry battle the IP pirates

Faced with the economic equivalent of a life-threatening situation, the media industries are fighting piracy on two fronts:

- with legal action to enforce intellectual property protection in countries where there are known patterns of copyright infringement; and
- through digital rights management (DRM) technology, which embeds code in the music or movie to impede copying, or to trace illegal copies back to their source.

On the legal front, governments are increasingly active in attacking piracy and pirates. A Hong Kong judge condemned a man to three months in jail for using an Internet file sharing program to make three Hollywood movies available for free downloads.[26]

The government of China announced in March 2006 that its supreme court had named a special Judicial Court of Intellectual Property to handle cases of illegal copying of copyrighted products.[27]

Amid such promising signs it is important to remember that the strict enforcement of intellectual property laws is still relatively rare in many of the world's countries. Without such enforcement and the prosecution of offenders, it is impossible to deter piracy.

Rather than wait for the legal climate to improve, some media companies are turning to more pragmatic ways of pre-empting piracy. Movie studios are making DVDs of their films available for sale much

---

[26] K. Bradsher, "In Hong Kong, a jail sentence for online file-sharing," *The New York Times* (November 8, 2005), Section C, 4.

[27] D. Visser, "China creates court for piracy cases," www.washingtonpost.com/wp-dyn/content/article/2006/03/10/AR2006031000275.html (accessed March 10, 2006).

earlier than before. In the case of video games, Electronic Arts is no longer offering games from its Chinese studios on disks, but only as Internet downloads. Games in this format are harder to copy than a disk.[28]

On the technical side, DRM remedies come in two forms:

1. Encryption or similar schemes to deter illicit copies of files.
2. Watermarking through which illicit copies can be traced back to their source.

### Encryption

Encryption is only marginally effective, because it can be broken by clever hackers, just the sort of people who might engage in piracy. What's more, it risks alienating the industry's honest customers. People who have paid good money for a CD, for example, are unhappy when copy protection prevents it from working in high-end CD players or PCs.

In their anti-piracy zeal, some media companies have gone beyond copy protection to use DRM schemes that could be considered an invasion of their customers' privacy. For some time, Sony BMG, one of the world's largest recorded music companies, was putting copy protection programs on their music CDs that installed hidden software on the hard disk of any PC used to play it.

The software (either XTC from First 4 Internet or MediaMax from SunComm Technologies, Inc.) then sent information from the PC to Sony without the customer's knowledge. Inadvertently it also made the PC vulnerable to attacks or takeovers by hackers. When the ploy was revealed to the public, customers were predictably outraged and Sony BMG found itself facing a public relations disaster and a spate of class-action lawsuits.[29]

### Watermarking

Watermarking, which incorporates invisible digital identifying marks in the video, is difficult if not impossible to break. However, it is a forensic tool, not a piracy preventive. It is useful in prosecuting

---

[28] G. A. Fowler and J. Dean, "Media counter piracy in China in new ways," *The Wall Street Journal* (September 28, 2005), B1.

[29] The Electronic Frontier Foundation, a group of consumer advocates opposed to many forms of DRM, maintains a website (www.eff.org) documenting such cases and their involvement.

offenders because it can identify the last legal owner of a bootleg movie. That person is likely to have been the pirates' source.

Another technological curse hanging over the movie and music industries is cheap software (i.e., peer-to-peer) for file serving over the Internet. This software allows the distribution of video and music files without the need to go back to a source, and thus without paying fees to the copyright owners.

The case of *MGM* v. *Grokster* shows how seriously content providers are taking this threat. The *Grokster* case reached the US Supreme Court in 2005. The issue to be decided was whether the provider of the free file exchange software, Grokster, was an accessory to the illegal distribution of music, and as such could be put out of business.

The court ruled that Grokster was protected by a prior legal decision (*Sony* vs. *Universal*, 1984). In this case the court ruled that technology developers who know their products are capable of infringement are not liable for illegal use as long as lawful uses for the technology also exist.

However, the Court left the decision open to review by a lower court if Grokster was found to have actively promoted the use of its software for copyright infringement purposes. In 2005, another file sharing company, KaZaA was ordered by a judge in Australia to change its software to prevent its use in copyright violations.

## Print in the digital age

Up to now we have been talking about the illegal distribution of copyrighted songs, games, and movies. But what about books still under copyright?

So far book publishers have been insulated from digital piracy, real or perceived, because electronic scanning of books was costly and the demand for any one title hardly justified the investment for pirates. This is changing, however. High-speed scanning has made it possible to turn books into electronic objects even if there are no original computer typesetting files available.

All that was missing was an organization willing to bear the cost and effort involved in the scanning process. In 2005 Google stepped forward, announcing that it plans to digitally scan entire libraries at selected major universities and make their contents available on line, in fully searchable form. The plan even made business sense. After all,

Google has become an economic powerhouse by supplying visitors with free information.

Google's initial proposal was to scan the books regardless of copyright status. Protests by the American Association of Publishers are forcing a change in those plans. Under copyright law, Google (and others doing similar scanning) could be fined $150,000 per violation, and the prospect of paying huge fines is not appealing. As a result, publishers will have the opportunity to opt out of the scanning program if they so choose.[30]

Other programs for digitizing books, such as that proposed by Yahoo!, will include works not covered by copyright, either because they are in the public domain or because their owners have given permission.

This is just the beginning. The process of digitizing printed media will make the protection of copyrighted books ever more difficult.

### Governments vs. IP holders

Obviously, companies that own valuable IP have good reason to be concerned. Technology is outstripping their ability to control the distribution of content. If the flood of piracy is not stemmed, they could face steep revenue declines as cheap pirated music, movies, and software kill the sales of legally-produced material.

As if the industry did not have enough to worry about, some governments are now on the offensive against accepted industry practices. Microsoft, for example, has been fined by South Korea in an antitrust case for bundling Media Player, a program to play audio and video material, and Microsoft Messenger, an instant messaging program, into the Windows operating system.

The Korean government says that local vendors with similar solutions cannot sell against such a dominant supplier, especially when that supplier is providing its software free. The chairman of the Korean Fair Trade Commission, Kang Chul-kyu, said that the bundling of software is an abuse of the market, and is "blocking competition and leading to a monopoly in the market . . . as well as raising the entry barrier to PC makers . . . and hurting the interests of consumers."[31]

---

[30] T. G. Donlan, "Go-go Google," *Barron's* (August 22, 2005), 39.
[31] A. Fifield, "Software group fined $31 m by Korean watchdog," *Financial Times, London Edition* (December 8, 2005), 32.

The European Commission also fined Microsoft $594 million (€497 million) in 2004 on similar grounds. In March 2007, Microsoft accused the commission of colluding with its rivals in pursuing a settlement of the case, a charge the commission denied.[32]

## Global questions

What are IP owners to do? Trade barriers are rarely effective in protecting them against foreign competition. Outsourcing and joint ventures keep them competitive for the time being, but the exporting of jobs has political implications, and locating key facilities in foreign countries can diminish their control of their own technology.

Even legal remedies such as patents and copyright don't ward off pirates, who have sophisticated technology of their own. And foreign governments will act to protect their own industries from domination, as South Korea and the EU are doing.

Globalization produces winners and losers, and in many cases technology is not the deciding factor. Managing contradictory trends is the key. As companies move from one strategy to another to protect and expand their positions, and as governments try to establish conditions for sustained growth for their most promising industries, only the flexible enterprise will survive.

This is what it means to live in interesting times. The global economy is being transformed by digital technology. The same technology that has enabled new models of communication, entertainment, and commerce is threatening the very existence of some of the leading companies in the field.

In the midst of these contradictions, government and industry are matched in an uneasy alliance, trying to control the forces that have been unleashed. We cannot know the end of the process – perhaps there is no end – but one thing is certain. They must work together, not just to hold chaos at bay, but to build sustainable, prosperous economies.

---

[32] For a summary see S. Taylor, "EU hits back at Microsoft complaints," www.computerworld.com (March 10, 2006) (accessed on March 12, 2006).

# 8 | *The digital world: Industries transformed*

I N just fifty years digital electronics have transfigured every aspect of human life and redrawn the economic map of the world. Their effect has been revolutionary.

But revolutions produce displacements as well. We must now assess the impact of digital technology on some key industries and the consumers they serve. We will discuss the implications for national economies in Chapter 9, keeping in mind that changes are so rapid that conclusions are risky.

We began our exploration of digital electronics by defining the breakthroughs of the 1950s and 1960s: the devices, systems, and software that launched the digital revolution. We also looked at how scientists and entrepreneurs created and commercialized the resulting products and services that grew out of these innovations.

Their efforts changed the world at incredible speed, often in unanticipated ways, and with far greater effect than even the most optimistic visionaries could have foreseen. Since the 1970s powerful digital systems have transformed the way people work, talk on the phone, send messages, acquire knowledge, play games, access entertainment, and even cure diseases. This is especially true of the Internet, the ultimate embodiment of the digital technology's power and reach.

Nor have the benefits accrued only to wealthy nations. Electronic devices are now so inexpensive that hundreds of millions of consumers in developing countries use cell phones, PCs, and MP3 players. Just two or three decades ago many of these people would have considered a standard telephone to be an unattainable luxury.

Vast demand from consumers and businesses creates huge opportunities on the supply side. As we have seen, corporate boardrooms and government offices alike have moved to exploit the potential of digital innovation for creating wealth and jump-starting economic growth.

We have seen how its emergence has raised serious issues for business and government leaders in four areas:

- The transfer of technical knowledge to developing economies.
- Venture capital stepping in to replace corporate support of innovation activities.
- The threat that offshore manufacturing poses to developed economies, even while it helps developing countries move toward prosperity.
- Attempts by governments to protect high-tech industries from foreign competition.

We will now consider the broader implications of these developments.

## Anatomy of change: Looking at fundamental technology markets

Much of the digital remaking of our social and economic life has come from four crucial areas of innovation. They are the Internet; the communications industry; the electronics industry (especially its impact on manufacturing); and the software industry.

We will not revisit the software industry, which was covered at some length in Chapter 2, because much of its transformational value is embedded in electronic systems. Besides, the breakthroughs driving the field's economic growth, such as open source software and software as a service (SaaS), could not exist without the Internet, which we consider in depth below.

For that reason we will focus on the Internet and its impact, the big changes in communications, and the global expansion of the electronics industry. In these three areas we will clearly see the balance of economic power shifting within industries and among nations.

This is no longer a world in which a few dominant providers in a handful of countries can monopolize production and the creation of wealth. The rules are changing and everyone, from individuals to governments, must deal with them. We will explore the new realities of the digital world as they affect the economic, social, and political spheres.

Our itinerary begins with the Internet, an innovation that musters all of the attributes of digital technology to bring people and organizations together as a global community. Then we will survey the turmoil in the communications industry, which not incidentally provides the underlying technology for the Internet.

Finally, we will scrutinize the electronics industry, which develops and builds digital systems. Readers who believe the whole world is evolving toward a service economy may be surprised at how important manufacturing still is.

## The Internet: Crosslinking the digital universe

The Internet is one of the most revolutionary innovations in human history. It is the convergence of remarkable new electronic systems with extraordinary advances in communications, and as such, embodies their transformational power.

Indeed, if the p-n junction was the electronic equivalent of the "big bang," the cosmic event that created the materials of the universe, the Internet is the digital ecosystem that grew out of that explosion.

Like any ecosystem, the Internet provides an environment in which single organisms and whole species can develop, interact, proliferate – and become extinct. The life forms of the Internet are individuals, companies, and large social structures, including governments. Their success or failure depends on how well they adapt to the environment.

Information is the digital ecosystem's life force. To move from metaphor to a functional description, we can think of the Internet, including the World Wide Web, as the pivotal contribution of electronic technology to the dissemination of information.

### *Surpassing print*

To a significant degree the Internet is an outgrowth of the printed word. Print and the Internet have a commonality of purpose: accelerating and broadening the spread of information. But a comparison of the two media also shows how different they are, both in what they do, and in the extent of their influence.

Gutenberg's brainchild revolutionized the world's social and economic structures, just as the Internet is doing in our own day. Before printing, recorded knowledge and learning were the private preserve of the privileged and educated classes. Printing broke new ground by making knowledge accessible to the general public.

For the first time in history people could easily read the Bible, for example, instead of having its messages filtered through the sermons of the ecclesiastical hierarchy. Through mass-produced books, broadsides, and periodicals the ordinary person could encounter information directly, instead of being dependent on hearsay, oral tradition, and the limited number of intermediaries who could read manuscripts.

In short, by democratizing knowledge and information, printing empowered its audience. Abraham Lincoln memorably characterized its unchaining of human potential. " 'When printing was first invented,' Lincoln [wrote], 'the great mass of men ... were utterly unconscious that their conditions, or their minds, were capable of improvement.' To liberate 'the mind from this false and under estimate [sic] of itself, is the great task which printing came into the world to perform.' "[1]

But printing has its limitations. It is not instantaneous, so the information it provides may not be current. It provides little opportunity for readers to interact with other readers. Each document is a separate physical product, which readers must obtain from specialist suppliers.

Most significantly, the production of printed materials requires expensive resources – presses, ink, paper, skilled labor. This limits the process to publishers who can afford them, and makes printing vulnerable to control by totalitarian governments. In A. J. Liebling's famous observation, "freedom of the press is guaranteed only to those who own one."

With the advent of the Internet, those constraints have almost vanished. Information gets posted to the Web as soon as it's available, so it's always up to date. E-mails, chat rooms, comment pages, personal and corporate Web sites, and blogs provide unprecedented opportunity for interaction among users. The whole experience is virtual: there are no books to buy, CDs to play, newspapers to recycle.

Last but not least, the users have ultimate control. While governments can limit access to the Web, it's much harder to do than it is to regulate printers or shut down a newspaper.

Both printing and the Internet opened new vistas in information access. But by providing services that were not practical before it came on the scene, it is the Internet that has fostered the creation of new communities, and new possibilities for collaboration. The Internet is changing our personal lives, and creating new business models while tearing down old ones.

The new applications constantly emerging for the Internet are bringing about fundamental changes in a wide range of economic and social

---

[1] D. K. Goodwin, *Team of rivals: The political genius of Abraham Lincoln* (New York: Simon and Shuster, 2005), p. 51.

activities. Our discussion will frequently refer to articles on these applications published in the general press. I have chosen them because they provide concrete examples that are indicative of broader trends.

## Internet search: Empowering the Web

One of the new business models generated by the Internet is the function commonly known as "search." Search is the service that makes the World Wide Web (and therefore the Internet) readily navigable by every user. Without it the Web would be a huge jumble of undifferentiated documents, difficult to navigate and impossible to use.

Search uses specialized, astonishingly sophisticated software to catalog hundreds of millions of Web sites. When a user asks for information on virtually any topic of interest, the search engine presents a list of links to sites with related content.

In effect we are talking about tapping vast reservoirs of intelligence distributed around the globe, and harnessing it for the purposes specified by individual users. This is unprecedented in the history of the world, and the results of its success are just beginning to emerge.[2]

Because of the importance of search, it is the logical starting point for a survey of the implications of the Internet and the Web. It also offers a case study in how an innovative approach to information technology can have effects that resonate far beyond its source.

### From research desks to electronic queries

Search did not spring full-blown from the minds of electronic engineers. It is analogous to earlier document-based research methods. That stands to reason, since the World Wide Web is at heart a vast collection of electronic documents.

In the days before the Internet, when a trip to the local library was the best way to pursue a subject, information retrieval was the specialty of expert research librarians. Instead of browsing the stacks, a library user could submit a query to a librarian.

---

[2] M. Meeker, B. Pitz, B. Fitzgerald, and R. Ji of Morgan Stanley Research, "Internet trends," a talk presented at *Web 2.0* (October 6, 2005). Accessed at www. morganstanley.com/institutional/techresearch/internet_trends_web20.html?page= research (accessed on May 9, 2006).

The librarian would use reference books, indices, and catalogs to find out what information on the subject was available on and off site in perhaps millions of books, journals, or other documents. Then the librarian would collect relevant documents for the user.

The Web, like the library, is designed for document sharing, but it goes about the task differently. Users could simply browse through Web sites they know about, as if they were browsing through the books on library shelves. But that approach would be time-consuming and inefficient, and they would certainly miss valuable information on sites they did not access.

To make search faster and more comprehensive, the developers of the Web looked beyond browsing. They created new procedures and protocols for searching across all sites, and for returning results with hyperlinks. The hyperlinks let users go to the documents without knowing where they are.

### Beyond cataloging

The Web's beginnings were humble. There were only 100,000 accessible text pages in 1994, and users seeking information submitted only a few thousand queries a day.[3]

Nobody could have known that a little over a decade later there would be over eleven billion pages of information available on the Web.[4] And they certainly never dreamed it would be possible to "catalog" and search all of these documents and answer specific information queries in seconds.[5] But that is exactly what search does.

Obviously the number of documents available on the Web today dwarfs what the research librarian could access in print. We are dealing with billions of indexed digital documents stored on magnetic disks controlled by hundreds of thousands of servers (computers); this takes powerful and highly sophisticated search engines.

---

[3] O. A. McBryan, "GENVL and WWWW: Tools for taming the Web," presented at *First International Conference on the World Wide Web* (Geneva: CERN, May 22–27, 1994).

[4] A. Gulli and A. Signorini, "The indexable Web is more than 11.5 billion pages," a research paper posted at www.cs.uiowa.edu/~asignori/web-size/ (accessed on March 27, 2006).

[5] F. Cowan, M. L. Nelson, and X. Liu, "Search engine coverage of the OIA-PMH corpus," *IEEE Internet Computing* (March/April 2006), 66–73.

Every day search engines respond to hundreds of millions of keyword-based queries. They answer each query in a matter of seconds. The secret is in the complex mathematical algorithms that rank documents by their anticipated value in answering a specific query. These algorithms are continuously updated to keep pace with the growing volume of documents and queries.[6]

A search engine also handles a pervasive problem that a librarian would rarely, if ever, face with printed materials. Documents on the Web come from uncontrolled sources, and since they are in electronic form they are easy to edit. As a result, they are in a constant state of flux.

When librarians catalog and shelve books they don't have to be concerned that the content between the covers is going to change the minute their back is turned, but that is essentially what happens on the Web. A search engine can't take a single electronic glance at the title and contents of the document (site name) and index it forever. The engine must revisit the site periodically and look not just at its title page, but also at the present content of its documents as revealed by standardized labeling techniques.

### Web gateways: Google and search technology

About 38 per cent of the Americans who go on line do so for search.[7] This is a testimonial to the importance of search for the Internet. Because search is so central, the companies in control of the best search engines have become the gatekeepers of the World Wide Web. In the process they have turned information retrieval into a valuable industry in its own right.

Their success is built around a new class of software architecture that lets them transform the huge amount of data on the Web into useful information. Efficient implementation of this software requires powerful and highly creative information management techniques.

There are many commercial search engines, but the clear leader is Google.[8] Its superior search results have made it far and away the most popular site of its kind, handling half of all queries on the Web. Its

---

[6] For a glimpse into the world of Internet search technology, see "Needles and haystacks," *IEEE Internet Computing* (May–June, 2005), 16–88.

[7] Pew Internet and American Live Project, cited in "The digit," *Newsweek* (February 27, 2006), 14.

[8] According to Google's Web site, the name comes from a deliberate misspelling of "googol," which is the fanciful name for a value of 10 to the power of 100. Some suspect it is also a pun on "googly," an old slang adjective for protruding eyes

success is a validation of its stated mission: "to organize the world's information and make it universally accessible and useful."[9]

On a technical level, Google's success derives from two advantages. First, its engine has enormous scaling capabilities. More crucially, it has developed very efficient methods for ranking documents in order of their predicted relevance to users' needs.

Google's ranking algorithm, called PageRank™, is the foundation of its superior performance. PageRank, developed by Google founders Sergey Brin and Larry Page[10] while students at Stanford University, works by looking at how many documents on other Web sites link to the page being ranked, and by determining the legitimacy of those links.

A simple way to characterize this process is to refer once again to the print world and think of a comparable situation in doing research with paper documents. If you run across a source that seems relevant, how do you determine its importance to your study? One way is to note how many other publications refer to the source as you continue to gather information. The more frequently others refer to it, the more important it is.

This is overly simplistic – PageRank factors in many site attributes beyond the raw numbers of crosslinks. Among other things, it looks at the quality and standing of the linking sites. But crosslinks are still at the heart of its rankings.

### Google: Search as a business plan

Entrepreneurs realized early on that document sharing was a key Internet capability. Many start-ups launched their businesses to exploit this capability by catering to interest groups that they wanted to attract to their Web sites.

They discovered that having an efficient search engine is not enough to build a viable Internet business. Monetizing the search service proved

---

(bugged-out, searching eyes have occasionally appeared in the two letter Os of the company's logo).

[9] Quoted from Google's corporate information page at www.google.com (accessed April 24, 2006).

[10] Their research was funded by the National Science Foundation, the Defense Advanced Research Agency, and the National Aeronautics and Space Administration. For an early description of their methods see S. Brin and L. Page, "The anatomy of a large-scale hypertextual Web search engine," *Proceedings of the Seventh International World Wide Web Conference* (Brisbane, Australia: April 14–18, 1998), 107–117.

very difficult. There was no problem in getting visitors onto the sites. What was missing was a model for generating revenues to pay for site development and support.

Sites tried to finance their operations by selling "banner ads" on their Web pages to generate revenues. This approach mimics the same ad-supported model used by specialty magazines. For example, a company called EarthWeb that catered to software developers had Web sites featuring ads for software products.

Unfortunately, ad revenues proved disappointing as a source of income. The problem was that advertisers could not determine the impact of their ads. They were understandably reluctant to continue paying for uncertain results.

The metric most commonly used to measure advertising results was a count of the number of visitors who clicked onto the site where an ad was posted. Since visitors could not click through to the advertisers' Web sites, advertisers could not judge the effectiveness of their ads in generating leads or sales.

A click metric is still used today but in a manner that is meaningful to advertisers. Google has to be credited for developing the approach that changed the rules of the game, and became the most successful business model for search.[11]

Google recognized that advertisers will pay more for search engine advertising if they can measure the impact of their ads. Accordingly, Google links ads on its site to specific keyword search results.

It works like this. In answer to a user's query for information on, say, "dc motors," the Web site will bring up pages of results listing free documents available on the Web. The documents are ranked in order of predicted relevance down the left-hand side.

At the top of the list on each page are ads, called "sponsored links," that look like search results. These appear because of their relevance to the keywords used in the search. Advertisers "buy" keywords, meaning they pay to have their ads served to a page in response to a query on those words. They compete on price for the placement of their ad on the page. Advertisers are billed at that price for the number of times (clicks) their sites are accessed.

---

[11] For an excellent introduction to the business behind Internet search see J. Battelle, *The search: How Google and its rivals rewrote the rules of business and transformed our culture* (New York: Penguin Group, 2005).

On the same Web page, on the right-hand side, the searcher finds other advertisements for companies selling dc motors. These, too, are called sponsored links. Typically, the higher the page position, the higher the price per click paid by the advertiser.

This approach, and variants from Google competitors, has been highly successful for search companies and advertisers alike. Online advertising revenues in the US in 2005 totaled $12.9 billion, and search engine companies had a 41 percent share.[12]

Advertisers have flocked to the search sites because they deliver much higher click-through rates. This is only logical, because the search engines are serving ads to people who requested the information they contain.

Google also encourages other companies to populate their Web sites with relevant ads served from Google. Its AdSense program pays participants every time a visitor clicks on one of these ads. In some cases it pays based on the number of people who see the ad when they log onto the Web site. About half of the revenues that Google derives from its advertisers are passed along to "partners" that provide links on their Web sites.

Letting advertisers track and pay for the number of viewer clicks on their ads accordingly seems like a bulletproof business model. However, some scattered but troubling reports of "click fraud" have recently appeared.

The fraud is based on the fact that advertisers generally limit their ads to a certain number of clicks per day, for example, to hold down expenses. Once an ad reaches that number of clicks, it gets pulled from the rotation.

Advertisers have detected an increasing amount of phony automated clicking on their ads. This is carried out by fraud specialists on behalf of competing merchants, who are trying to get competitive ads pulled early in the day. If they succeed, their ads get more exposure to potential buyers throughout the day. Meanwhile the advertiser being victimized pays for wasted clicks and loses sales opportunities.

If merchants don't get the value they pay for, the whole online advertising industry could be the loser. It remains to be seen how big

[12] Cited by eMarketer in an April 25, 2006 Web site article accessed at www.emarketer.com; figures are based on analyses by the Internet Advertising Bureau and PricewaterhouseCoopers.

the problem really is, but its increasing incidence has been flagged in the press.[13] The online advertising industry is taking the threat seriously and looking for remedies.

### Web 2.0 expands search horizons

The Internet emerged during the late 1990s, but the true measure of its impact has only been felt since the early 2000s. This period has been called the Web 2.0 Era, to indicate that the Internet is in its second generation.[14]

As part of this new direction, what started as a search capability is rapidly expanding into other services. The dominant search companies are offering telephone service, online data storage, and e-mail service. They are even hosting software as a service (SaaS), in competition with established software vendors.

They are riding a larger trend. After all, Web 2.0 already hosts such amenities as social networking sites, podcasts, blogs, VoIP, music, games, software services, and video. But search companies haven't forgotten their roots. They have also conceived new search ideas, such as "intelligent search agents," to find information specifically tailored to an individual. These agents can "learn" from past searches, as well as from user profiles, to continuously improve the quality of search results in a given area of information.[15]

Given the colossal clutter of the Web, search is indispensable. Without it users simply could not find their way among the geometrically expanding store of information making its way on line. Search lets them identify, locate, and research information, products, or services they need.

The search companies with the largest "libraries" of Web links are the guardians of, and guides to, the Internet. As long as they expand their libraries and build partnerships to disseminate information,

---

[13] T. G. Donlan, "Truth in advertising," *Barron's* (January 30, 2006), 51. M. Veverka, "How click fraud just got scammier," *Barron's* (June 19, 2006), 44.
[14] R. O'Reilly, "What is Web 2.0: Design patterns and business models for the generation of software," (accessed on www.oreilly.com on October 30, 2005).
[15] Surveys of this technology can be found in A. Levy and D. Weld, "Intelligent Internet systems," *Artificial Intelligence* 11 (2000), 1–14; and in A. Ferreira and J. Atkinson, "Intelligent search agents using Web-driven natural-language explanatory dialogs," *Computer* (October 2005), 44–52.

they will continue to be the first place users go to find it. That has enormous significance for them, their advertisers, and the whole online universe.

## The Internet generation (and others) do everything on line

In a single decade, an amazingly large segment of the public has embraced the Internet. They have done so with a zeal that is redefining practically every aspect of commercial and social activity. They exploit what search makes possible, with effects we discuss below.

This rapid adoption of the Internet for personal and business activities can be explained by several factors.

- The increase in information-related jobs means a large and growing majority of the workforce in the US and other developed countries are now computer-literate. The percentage of workers employed in the information sector, who are presumably comfortable with computers, increased from 36.8 per cent in 1950 to 58.9 per cent in 2000 and is still rising.[16]

- Personal computers are affordable. As a result, market penetration in the US passed the 60 per cent mark in 2003.[17] Figure 8.1 shows that this is still an upward trend, no doubt spurred by the fact that major makers began selling entry-level PCs at prices in the $400 range in 2005.[18]

- Access to the Internet is readily available in most parts of the world. An estimated billion people can go on line with dial-up connections through their phone lines. China alone had over 94 million Internet users in mid-2005. Broadband subscribers, those who get service at over 1 MB/s, topped 180 million in 2005.[19] With access now possible via wireless as well as wireline connections, the Internet could soon be available anywhere.

---

[16] E. Wolf, "The growth of information workers in the US economy," *Communications of the ACH* (October, 2005), 37–42.

[17] Data from the Consumer Electronics Association, reported on www.ebrain.org.

[18] From a presentation by D. Levitas, "The next frontier: Conquering the digital home," IDC (March 2005). Data collected from US Census Bureau, Statistical Abstracts of the United States, IDC, and vendor reports. This chart includes data from Consumer Electronics Association (CEA) Market Research.

[19] Meeker, Pitz, Fitzgerald, and Ji, "Internet trends," a talk.

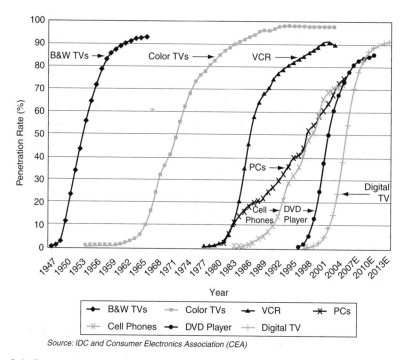

**8.1.** Penetration rate of some major consumer products in the US since 1947. Reprinted with permission from the IDC and the Consumer Electronics Association (CEA) (ref. 18).

- There have been massive investments in technologies enabling Internet services. Venture capital investments in new Internet-related start-up companies between 1994 and 2005 totaled about $200 billion, as cited in Chapter 5 (Table 5.1). VCs continue to fund Internet companies, but this is only part of the story. Established companies are building their own Internet capabilities with internal capital. Although many start-ups disappeared, there were a number that did succeed in developing technology and compelling online service and entertainment offerings.
- A large share of the credit for the dramatic increase in Internet usage by the general public must go to search engines. People love to search, and the easier it is to access and navigate the Web, the more attractive and valuable they find the online experience.

## The eclipse of print media

With search thriving and the world moving on line, print is becoming less relevant. Print industries are watching their influence wane and their revenues fall in an increasingly digital world.

Not surprisingly, a report by the Forrester Group indicates that the more people become accustomed to accessing the Web, the less interest they have in reading printed material.[20]

Proof can be found at that bastion of print, *The New York Times*. While the *Times* produces one million newspapers a day, its Internet news source, www.NYTimes.com, has over eighteen million distinct users monthly.[21]

Not surprisingly, the *Times* is committed to expanding its online presence. It bought the online portal About.com in 2005 to shore up its electronic offerings. In early 2006, Leonard Apcar, editor-in-chief of www.NYTimes.com, told an industry audience that although the print side of the business was downsizing, his division was 200 strong and growing fast.[22]

Apcar also pointed out a more interesting development. The *Times* is integrating its Web and print organizations, previously distinct operations, to leverage the content generation strengths of the old newsroom for the online product.

The reason: people who do not subscribe to a print edition seem to be willing to pay for online access to its contents. The company's newest online offering, a fee-based news and archive service called TimesSelect™, quickly enrolled 200,000 subscribers at $50 per year, in spite of the fact that users can get it free if they also sign up for the printed newspaper.

Editorial content is not the only kind of traditional newpaper content moving on line, of course. As eyeballs leave print media, so does advertising.

---

[20] C. M. Kelley and G. DeMoulin., "The Web cannibalizes media." Technical report by the Forrester Group, May 2002.

[21] D. Radev, J. Otterbacher, A. Winkel, and S. Blair-Goldensohn, "NewsInEssence: Summarizing online news topics," *Communications of the ACH* (October, 2005), 95–98.

[22] Remarks made at the BusinessWire briefing in New York on "New media technologies: Why it is vital for PR and IR pros to tune in before being tuned out," (March 21, 2006).

The increasing loss of local classified advertisements, which accounted for $15 billion in US news media revenues in 2005, is especially worrisome for newspapers. For example, craigslist.com offers free Internet classified listings in fifteen markets. Offerings like this cost papers in the San Francisco Bay area $60 million in lost job ads in 2005.

It is estimated that about 5 percent of US advertising revenues in 2005 were channeled through the Internet. As noted earlier, that amounts to $12 billion worth of online advertising, 41 percent of which went to Google.[23]

This phenomenon, which comes at the expense of broadcast and print advertising, is not exclusive to the US. Online advertisement revenues in the UK increased by 62 percent in the first half of 2005.[24]

Success on this scale attracts more investment, and more innovation. Online ads are expected to grow 15 percent a year for the next few years. Meanwhile, ad agencies and Web sites are seeking more effective ways to present advertiser messages.[25]

### GlobalSpec: Replacing catalogs with clicks

There are precedents for established print publications simply disappearing under an Internet-based onslaught. One of the most suggestive examples is found in the business-to-business process of sourcing industrial products and services.

For over 100 years, generations of US engineers relied on the *Thomas Register of American Manufacturers*, an annual Yellow Pages-like set of very thick books that provided information on tens of thousands of types of industrial components. A full shelf was needed to store them. Finding the manufacturer of a component such as a resistor, for example, meant searching the pages of the *Register* for ads and listings, and then contacting the vendors for specifications and prices.

The *Register* was so important to suppliers that it was not unusual for companies to allot their entire advertising budget to placements in every product category where they had an offering. This was before GlobalSpec, a start-up founded in 1996 (with Warburg Pincus funding)

[23] S. Lohr, "Just Googling it is striking fear into companies," *The New York Times* (November 6, 2005), 1.
[24] "Online advertising: Classified calamity," *The Economist* (November 19, 2005), 67.
[25] T. Ray, "Google faces stiffer competition as online advertising goes glitzy," *Barron's* (November 21, 2005), 32.

by former General Electric employees John Schneiter and Thomas Brownell, built a novel Internet product search capability.

Schneiter and Brownell had the idea of hosting product catalogs on line. They designed a search engine that could quickly find a desired product on the basis of its specifications, not just its name, and then provide links to the Web sites of the participating vendors.

For example, if an engineer enters the keyword "carbon resistor" on the GlobalSpec site, a Web page appears with links to information about resistors from publications and standards organizations. There are tabbed sections for searching according to products, part numbers, properties, and other criteria. And of course there are the links to vendors' Web sites. The vendors pay GlobalSpec to list their catalogs.

To attract engineers to its site, the company has over 250 million technical documents accessible to registered users as a free service. The more than two million registered GlobalSpec users can also elect to receive periodic e-mails about new products in their selected categories.

The company does not ignore the big search engines. Google, Yahoo! and MSN act as gateways to the site, popping up links to GlobalSpec when users search on any of 600,000 keywords. Catalog owners who pay for listings on GlobalSpec benefit from the extra traffic this brings.

By any measure, GlobalSpec is a success. At the start of 2006 it could search over 17,000 vendor catalogs, and give its two million users access to 76,000 suppliers. That's a huge advance over printed directories.

From the user's perspective, however, it is the intrinsic efficiency of GlobalSpec's online sourcing process that makes it a successful endeavor. Because of the quality of the site's search results and its comprehensive catalog list, engineers can source a new product in minutes. With print catalogs this would take days or even weeks.

What happened to the *Thomas Register*? It went on line in the late 1990s with a search function to access the products in its lists. Meanwhile the print edition shrank from thirty-three volumes in 1997 to twenty in 2005, its last year. The *Register* books are no longer printed and will not come back.

**Universal online access to broadband content: Voice, data, video**

The Internet, as we learned in Chapter 2, does not discriminate among different types of content, provided they are digital and packetized. Some packets need special handling, but in general all content is transmitted the same way.

As a result, the Internet is as adept with voice and video as with documents and data. All that's required is enough bandwidth to handle the huge numbers of packets that make up, say, a one-hour video presentation.

That requirement is being met. Broadband availability is expanding rapidly, giving consumers around the world convenient online access to voice, music, movies, and games. The Morgan Stanley report cited above, which put the total number of broadband subscribers worldwide at over 180 million, shows that the trend has gone global.[26]

For example, the report points out, the percentage of homes in the Netherlands with broadband access reached 55 percent in 2005. The US currently trails, with only 32 percent, but that number is expected to nearly double by 2011. South Korea has probably the highest broadband penetration in the world, exceeding 70 percent.

Consumers consider broadband access to be only half the value equation. They're not interested in sitting down at a PC and logging into the Internet every time they want to hear a tune or see a video. They want to capture the content and use it on their own terms.

And in fact, the availability of low-cost hard disks (magnetic storage, discussed in Chapter 1) makes capturing this content practical. As Figure 8.2 shows, a 160 GB disk (about $100 in 2006) can hold 90 hours of video and 2,700 hours of music.[27] Consumers can keep the content on the disk, transfer it to other media for portability, or stream it to a playback device in another part of the house.

Consumer capture and storage represent a seismic shift for the media business.

### The Internet and cheap digital storage: Challenges for the media industries

From the days of Edison, the media industries have made their money by selling entertainment through controlled channels to mass audiences.

The media maximize their profits from music, movies, and television programming by segmenting distribution geographically and by market channel. To take a contemporary example, when a movie has finished its theatrical run in the US, it goes into distribution overseas,

---

[26] Meeker, Pitz, Fitzgerald, and Ji, "Internet trends," a talk.
[27] Data from Parks Associates (2005).

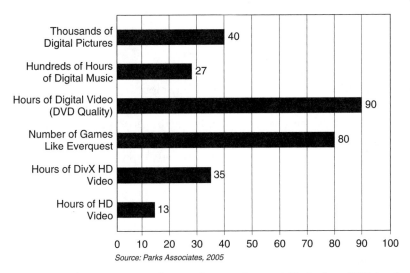

**8.2.** Entertainment content that can be stored on a 160 gigabyte (GB) hard drive. Reprinted with permission from Parks Associates (ref. 27).

while the studio issues a DVD of the movie for domestic purchase or rental.

The disk is regionally coded so that it will not play on DVD machines outside the US. This is done to keep unauthorized disk sales into foreign countries from cannibalizing overseas theater revenues. When a movie finishes its run in an overseas region, the studio issues disks coded for that market.

Studios also sell the movie to premium cable TV channels after its theatrical run. Eventually it is sold to network TV. Reuse of content is obviously the cornerstone of this business model.

But the model is breaking down. Two major developments are reducing the demand for video DVDs and prerecorded music, and changing the value proposition for broadcast TV as well. Both grow out of the power of the Internet.

First, as we have just seen, the spread of broadband makes it possible to distribute video and music directly to consumers over the Internet. Second, Internet availability of TV shows enables a highly sophisticated form of on-demand viewing (i.e., time-shifting).

We'll start with the sale of content over the Internet. This was pioneered by Apple and its iTunes music download site. Geared to owners of the iPod (Apple's portable playback device), iTunes has been extremely successful. At this writing it accounts for 80 percent of all digital music sales over the Internet.[28]

Video distribution is ramping up as well. Over twenty-five million US Internet users visited sites offering downloadable videos in January 2006.[29] Big search engines such as Google and Yahoo! are major entry points for consumers seeking portals where video, music, and related content is stored.

The growth and popularity of such portals happened spontaneously rather than by plan. Once established, however, they demonstrated significant value as a means of accessing special groups of customers. For example, when NBC Universal acquired iVillage (which bills itself as "The Internet For Women™") for $600 million in 2006, the portal already had 14.5 million unique users.[30]

So the audience is there, it can be segmented by special interests, and the facilities are in place. Web sites such as www.CinemaNow.com are offering legal downloads of Hollywood films on a rental or purchase basis, and promise that customers will eventually be able to transfer them to DVDs or portable video players, once the studios agree on security to protect the content from piracy.

While this looks like another potential market where studios can resell content, there is a sticking point: content protection, also called digital rights management (DRM). DRM is supposed to prevent savvy hackers from purchasing one legal copy of a movie, then duplicating it on inexpensive DVD disks to share with friends or sell for profit.

Unfortunately for the industry, no DRM scheme is completely hacker-proof. Even the best content protection is no guarantee that the movie will not be pirated.

Is developing an Internet distribution channel worth that risk? The music industry, which was the first to embrace digital distribution, has now had plenty of time to evaluate the security of the new medium, and so far the news is not good.

---

[28] B. Stone, "Can Amazon catch Apple," *Newsweek* (February 13, 2006), 46.

[29] Reported in S. McBride and M. Mangalindan, "Coming attractions: Downloadable movies from Amazon," *The Wall Street Journal* (March 10, 2006), B1.

[30] R. Morgan and C. Murphy, "NBC Universal takes iVillage for $600 million," *The Daily Deal* (March 2, 2006), 1.

In the several years since the digital music distribution channels launched, total worldwide music industry revenues (about \$22 billion at wholesale prices in 2005) have steadily declined. This is because sales of recorded music fell faster than Internet sales grew.[31] The industry has been quite vocal in its claims that unauthorized downloading and pirated CDs are the cause of the problem.

TV faces a similar dilemma. Broadcasters are concerned about the availability of video on demand, just as their counterparts in the movie and music industries are about post-sale duplication.

People are using their PCs to download television shows from the Internet, capture them off the air, or record them from a cable feed, and store them for later viewing. This undermines the basic premise of commercial TV. Advertisers rely on dedicated broadcast time slots to target audiences, and pay broadcasters accordingly.

Now viewers can shift the shows out of those slots. Worse, they can delete advertisements automatically, reducing the willingness of advertisers to finance programs.

As we discussed in Chapter 7, new technologies for controlling content are being developed, but it's clear that the industry will also seek regulatory protection to safeguard its revenues.[32] Whether these efforts will bear fruit is uncertain. What is certain is that the historical pattern of mass distribution of movies and TV programming is no longer the solid, secure business model it once was.

### Shopping is an Internet experience

No account of the impact of the Internet would be complete without mentioning consumer purchases. This is the engine that drives the economy, and it is the killer application for the Internet as well.

An incredible range of products is sold on line. Do you need a part for an old barbecue? A coffee cake, diamond ring, light bulb, computer, toy, house, classic car? Whatever you want, it's available. What's more, thanks to the quality of the search engines, you can find it in seconds from multiple vendors who are clamoring for your business.

---

[31] J. Leeds, "Music industry's sales post their sixth year of decline," *The New York Times* (April 1, 2006), C2.

[32] A. Schatz and B. Barnes, "To blunt the Web's impact, TV tries building online fences: Stations are using technology to restrict who watches; seeking aid from Congress," *The New York Times* (March 16, 2006), 1.

In a total shift in the balance of power between buyers and sellers, the buyers are in control. Using comprehensive information available free on line, they can rank one product against others and from multiple sellers and do a price comparison while they are at it. Companies and retailers who once relied on consumer ignorance to help them sell inferior products at premium prices are finding it much harder to play that game.

That does not mean that online merchants have no cards to play. Brick-and-mortar merchandisers use nice stores, pleasant salespeople, and extensive advertising to gain the advantage over their rivals. Online vendors get their competitive edge from superior Web sites to ease the shopping experience, the right search keywords to bring buyers to their sites, and high rankings for their ads on a Google Web page.

Here are some of the distinctive advantages that make Web-based stores a threat to conventional retailers, especially those who ignore the Internet as an adjunct to store sales.

- *Buying online is a global experience.* As Figure 8.3 shows, 40 percent of all Internet users around the world purchased merchandise online in 2004.[33] Studies show that over 90 percent of buyers were satisfied with their shopping experience.[34]

- *Online retailing can be more flexible.* Direct contact between vendors and consumers on the Internet makes it possible to sell customized products at standard-product prices and still be profitable. Dell built a very healthy computer business this way. The flexible, interactive nature of the medium promotes the creation of unique business models, too. The auction model invented by eBay is a prominent example.

- *Size and cost are no obstacles.* Although Internet retailing started with inexpensive items such as books, today even Amazon.com, the pioneer in this sector, is selling large, costly items over the Web. The buying public now has a high degree of confidence in the credibility of the vendors and the security of payment methods.[35]

---

[33] R. Peck, CFA, "Vistaprint Ltd: A paradigm shift in printing," Bear Stearns Equity Research (November 16, 2005), p. 26.

[34] Summarized by M. Meeker and R. Ji, "Thoughts on the Internet and China," Hua Yuan 2005 Annual Conference (June 7, 2005).

[35] M. Mangalindan, "Size doesn't matter anymore in online purchases: Web shoppers get comfortable buying big-ticket items; 3,600 pounds of copper gutter," *The Wall Street Journal* (March 22, 2006), D1.

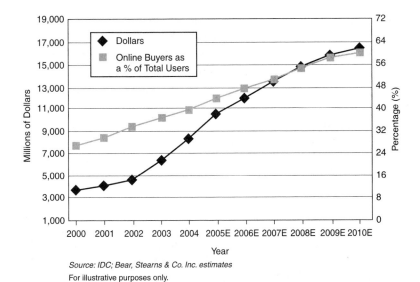

Source: IDC; Bear, Stearns & Co. Inc. estimates
For illustrative purposes only.

8.3. Global online commerce since 2000 showing the dollars of sales and the per cent of online users that shop on the Internet. Reprinted with permission from Bear Stearns (ref. 33).

- *Voice, data, and video are making Internet marketing more attractive.* To cite just one example, the real estate industry is using online animated maps and detailed, up-to-date local information to make the search for property faster and simpler. Using VoIP, customers can talk to an agent through a direct, free Internet phone call placed through their computer, while watching a video of the house or office under consideration.
- *Mom-and-pop retailers can compete with giant chains.* The big chains are getting a challenge from an unlikely source: the Web version of "neighborhood" stores. While the brand recognition and marketing muscle of a Wal-Mart or Best Buy still give them an advantage, it's not a decisive one in cyberspace. When people search for products, the results show the smaller retailers right alongside the larger ones.
- *The Internet can lower order fulfillment costs for retailers.* For example, Avon Products, the world's largest direct sales cosmetic company, replaced paper forms from its reps with Internet-based order fulfillment. This reduced processing costs from over a dollar

per order to a few cents each.[36] Another valuable role of the Internet in fostering commerce is in managing the flow of information that encompasses the product supply chain. This is a benefit exploited by the carriers such as Federal Express to great advantage.

Avon's strategy shows that old and new retailing models can mix. Several other US retailers have instituted hybrid approaches of their own, usually dubbed "clicks and mortar." One popular model is using physical stores as drop-off points for Internet orders and returns.

But there can be no doubt that, on balance, the virtual mall of the Internet is pulling shoppers away from actual stores, especially during the holidays, when customers want to avoid the crowds.

### Online social networks

Commerce is not the only arena where the Internet has changed the rules of the game. It also has a unique ability to foster social networks and host opinion exchange forums. Because it can accommodate and serve interest groups of any size, it can scale from providing a platform for personal expression to creating forums for the exchange of views and information at any level of organization.

There seems to be a Web site for every interest group, and it's not always about mass market appeal. Niches are everywhere on the Web. Their individual audiences may be small, but in the aggregate the numbers are huge. One example of a niche audience attraction is Sail.tv, a sailing "channel" created in the UK that covers sailboat races and offers videos on sailing techniques.[37]

Other sites cater to much larger groups. Teenagers, the group-oriented pioneers of all things on line, have endorsed the social aspect of the Internet almost unanimously. Surveys show that 87 per cent of teens use the Internet as a preferred place for socializing and purchasing products that cater to social fads.[38] The Web site mySpace.com, which serves as a social forum for more than 50 million young people, built its business model around their longing for a place of their own on the Web.

---

[36] Andrea Jung, Avon CEO, quoted in "Ideas outside shot: A conversation with Andrea Jung," *BusinessWeek* (February 6, 2006), 104.

[37] S. Hansell, "Much for the few: As Internet TV aims at niche audiences, the slivercast is born," *The New York Times* (March 12, 2006), BU1.

[38] J. Hempel, "The MySpace generation," *BusinessWeek* (December 12, 2005), 88–94.

Teens are hardly the only people forming Web-based bonds with like-minded spirits. Earlier we mentioned iVillage, the forum for women's issues accessed by more than 14.5 million unique users annually. However, mySpace.com hosts over three times that many young people. Media giant News Corporation acquired mySpace.com in July 2005 as the major property in a $580 million purchase of its parent company.

### Power to the wired people

Web site involvement in the political arena is also growing. It ranges from one-person initiatives to campaigns organized at the local, national, or even international levels. The most interesting of the individual efforts is the "blog."

Blogs (the name comes from "weB LOG," a diary of online activity) are easy-to-build Web sites that individuals use as platforms for their views. Blogs have become quite influential in shaping public opinion on a variety of subjects, from product quality to political issues.

There are tens of thousands of blogs, accessible to anyone with an Internet connection. The sites can be accessed through search engines, and there is software available to broadcast the latest updates to interested Internet users.

They have a surprisingly large audience. Morgan Stanley Research data suggest that 27 percent of US Internet users read blogs.[39] They are quoted in TV newscasts, and some bloggers have become minor celebrities.

At their most powerful, blogs harness "grass-roots" political opinions and bring pressure on politicians to take positions on issues of the day.[40] But they also have unexpected applications in the commercial world. Blogs written by company executives attract attention the company would not otherwise get. Blogs also provide a forum where consumers can air opinions on products and services. This creates a powerful new tool for gathering market information.

Politicians were slow to tap into the power of the Internet, but in the last few years it has been used for issues and election campaigns at every level of government. Political strategists have teamed with

---

[39] Meeker, Pitz, Fitzgerald, and Ji, "Internet trends," a talk.
[40] A. Cohen, "Bloggers at the gates: What was good for eBay should be good for politics," *The New York Times* (March 12, 2005), WK 11.

technologists to develop methods for reaching target audiences directly, including the latest fad, daily podcasts to send to the faithful.

We are talking about a platform for voter outreach that is replacing older communications methods such as television broadcasts to mass audiences. The rapid rise of Howard Dean in the 2004 US Democratic presidential primaries was attributed in large part to his canny use of the Internet to rally his supporters. To quote Ken Mehlman, the Republican national chairman, "The effect of the Internet on politics will be every bit as transformational as television was."[41]

It should not come as a surprise that some governments are unhappy with the freedom of expression possible on the Internet, and deliberately try to keep their citizens from accessing content they consider objectionable or inflammatory. While it is difficult to censor what people see on the Internet, it can be done. Governments can set up gateway servers to scan the packet traffic and filter out documents that contain certain words. In effect, search engines become tools for blocking access to certain hyperlinks and the Web sites they represent.

Fundamentally, however, this is a rearguard action. Some contraband content is sure to slip past the monitors. And even on "permitted" Web sites people are likely to encounter perspectives that make them reconsider what they hear from their government.

## Turbulent era in communications

Digital communication has been a profoundly revolutionary force for progress over the past two decades. It literally enabled global industrialization and the Internet. Without high-speed digital links the instant exchange of information, which is at the heart of international business today, would be impossible.

As we discussed in the first two chapters of this book, digital communication got its real commercial start in the 1970s with the maturity and convergence of five revolutionary innovations: computers, fiber optics, switched packet networks, semiconductor laser diodes, and integrated circuits. Mix these innovations in a potent technological brew financed by huge amounts of capital and extensive government

---

[41] A. Nagourney, "Internet injects sweeping change into US politics: Terrain shifts since 2004 as time-honored tools move onto the Web," *The New York Times* (April 2, 2006), 1, 23.

support, and then add the wireless technology of the 1990s, and you have our current state of ubiquitous communications.

This is the nervous system of the industrialized world, within which the Internet is taking an increasingly central position. And let us not forget that digital communications provides the infrastructure for the Internet.

You would think that the communications industry, having invested hundreds of billions of dollars to make this all possible, would now be reaping huge financial rewards. But that is not how things have worked out.

The first big wave of communications growth and investment, which took place during the 1990s, ended with the financial collapse of 2000. It is estimated that investors lost over \$2 trillion (US) dollars in the crash of public markets and the ensuing bankruptcies.

The collapse of the bubble took down many new companies, including over 250 newly created public carriers in the US. But they left behind an enormous global communications infrastructure which is just now exerting the full force of its influence.

Today the industry is dealing with the heritage of the freewheeling 1990s, facing challenges from new competitors on the one hand, and confronting vast opportunities on the other. We are now riding the second big wave in communications, which is once again affecting nearly all commercial activities around the world.

## Sources of unrest

Telecommunications used to be a quiet, predictable business. Its transformation, at least in the US, began with the 1984 breakup of AT&T, the regulated monopoly telephone company, into eight separate businesses: seven Regional Bell Operating Companies (RBOCs) plus AT&T as a long-distance firm and equipment provider. Bell Labs went along with AT&T.

That was the beginning of twenty years of corporate turmoil. In 1996 AT&T spun out its equipment business, along with two-thirds of Bell Labs, as Lucent Technologies. The smaller RBOCs were eventually bought up by the stronger ones, essentially reconstituting much of the old AT&T monopoly, but without any involvement from AT&T.

The latest chapter of the saga has now been written. Four former RBOCs have been reunited with AT&T under the original corporate name. This leaves two dominant wireline carriers in the US.

Before all this happened, the communications industry was made up of three distinct businesses: wireline data and telephony, cable TV, and wireless voice services. The same structure pertained, with regional variations, in countries outside the US. Governments provided benign supervision to ensure orderly markets where everybody could make an attractive return on investments.

That is no longer the case. With the breakup of AT&T in the US and deregulation in other countries, many new carriers have been able to enter the long distance and data transport markets since the 1980s. The result has been sharply declining long distance telephony and data transport prices.

At the local level the marketplace in the US became even more chaotic. The Telecommunications Act of 1996 opened local loop access to newcomers on a low-cost lease basis, ending the local service monopoly of the RBOCs. This event sparked the introduction of nationwide DSL broadband service over copper lines to consumers by independent companies such as Covad Communications, founded in 1997 with funding from Warburg Pincus. These DSL providers operated in competition with the telephone companies.

It was a watershed moment in communications history: the availability of broadband Internet access to telephone service subscribers.

## Competition among carriers

Deregulation exposed telecommunications carriers to competition, but it is technology that drove the process. The massive rollout of packet-switched digital communications using the Internet Protocol (IP) made it possible for aggressive companies to challenge market incumbents with attractive new services at affordable prices, while rapid advances in cellular wireless technology made cell phones a viable communications option.[42]

Cable TV operators are also competing against telephone companies for their voice business. By upgrading their networks to carry digital traffic in parallel with analog video, cable companies can offer subscribers voice service and broadband data along with television. Subscribers

---

[42] Note that this situation is not limited to the US. The communications industry finds itself dealing with similar issues all over the world. Big waves in technology do not respect national boundaries.

who sign up for this "triple play" package can get lower rates for telephone service than those available from the phone company.

To make matters worse for the phone companies, the pricing structure of voice over IP (VoIP) telephony also threatens to undermine their voice business. VoIP providers can afford to offer a flat per-month fee for all calls, because the cost to carry voice, video, or data is exactly the same. But switched-network telephone companies earn a lot of their revenues, and thus their profit, on voice calls metered by the minute. This puts them at a distinct disadvantage against VoIP telephony offered by competitors.

VoIP service is slowly eating into the revenues of telephone companies not just in the US, but around the world. In Germany, for example, subscribers to United Internet AG's broadband DSL service are offered a choice of VoIP plans: either a flat rate of € 10 a month for unlimited national calls, or metered calls at only € 0.01 per minute.

The free Skype computer-to-computer telephony service, with over 100 million customers around the world, offers a different but no less serious challenge. Other Internet companies are also offering VoIP service.

Competition is not limited to the wired networks. In one decade the cell phone has gone from luxury to necessity, with market penetration in some countries exceeding 100 percent.

One effect of this growth is that wireless phone "lines" are displacing wired lines, both switched and IP-based, all over the world. Every year the number of wired phone lines drops a few per cent as subscribers decide they don't need both forms of service and rely exclusively on cellular.

Cable companies are not immune to threats to their operating revenues. Their monopolies in local markets offer little protection from three new sources of competition.

- Telephone companies are responding to the cable encroachment into voice service with video service offerings on their own lines. By extending fiber optic networks to homes they can provide cable-like video content, including video-on-demand, via IPTV.
- Movie studios can sell viewers direct access to video content over the Internet, diverting entertainment revenues from the cable operators. Customers purchase a video and download it for storage on a PC or recordable DVD, or order it for limited viewing as a video on demand.

- Internet portals (Yahoo!, Google, etc.) now provide streaming video directly to subscribers, bypassing the traditional providers of entertainment content.

## The battle for revenues

The growth in Internet traffic, primarily due to the popularity of video and search, has increased the demand for backbone capacity on the networks.

Unfortunately, although the wireline and wireless carriers own the network's distribution "pipes," they do not get incremental revenues from the sale of what they deliver. The content owners and Internet companies that collect advertising revenues reap most of the profits.

To increase the pressure, ubiquitous access through the Internet and wireless devices is making subscribers less willing to pay for transport. The value proposition has shifted toward content owners. The carriers are in danger of becoming commodity "bit-providers" and losing their lucrative margins.

Cellular carriers, in particular, could use the revenue to support their tremendous expansion. In a matter of a decade they have blanketed the world with cellular wireless networks, and have launched at least three generations of service.

The new, packet-based third generation (3G) networks let mobile subscribers access the Internet for data, video, e-mail, and voice connections. This makes the cell phone the most versatile consumer access device available, replicating the functions of the PC, PDA, portable television, and telephone. A whole generation uses the cellular handset as its primary electronic device.

Having invested in their costly 3G network infrastructure, carriers need to increase their monthly subscription fee revenues to cover the cost. They are looking for new services that consumers are willing to pay for.

So far this has been a slow process. Voice calls still generate the bulk of the revenues. In fact, voice call revenues constitute, on average, between 80 and 90 percent of the total wireless revenues of carriers in the US and the EU. The other 10 to 20 percent comes from auxiliary services such as short digital messages (SMS) and ring tones.

Both wireless and wireline operators dread the prospect of remaining poorly compensated carriers of valuable content over their expensive broadband networks, especially as voice traffic revenues grow slowly in the developed economies. But how can they participate in the revenue stream? Once an Internet connection is opened on their network, a direct path between buyer and seller is also open. Creative billing is the only way that carriers can share in the profits generated by the bandwidth-hogging traffic they provide.[43]

Content providers are pressing their advantage by directly accessing consumers through their own branded cellular services. For example, Disney's service uses the facilities of Sprint Corporation.

## Carrier competition: Strategies and outcomes

In spite of work-arounds by content companies, the network operators – the carriers who own the pipes – still control customer access. They are working to solidify that position against the competition, and to find ways to make their businesses more profitable.

The carriers' struggle to increase market share will not be decided easily or quickly. The new broadband technologies can reach consumers in too many ways. But smart carriers know they have to start by choosing markets where their technology gives them the best chance for success. They must have the technology that gives them maximum flexibility and operational efficiency within their market segments, and provides high-quality services at low delivered cost.

Wireline carriers, to start with the largest group of providers, are focused on high-bandwidth data transport for enterprises. This is where their ownership of high-capacity, globally accessible networks gives them a distinct advantage. Cable operators and wireless carriers simply cannot compete in this market space right now.

That doesn't mean wireline carriers can sit pat with what they have. To sustain their lead they have to give their customers better broadband access by bringing fiber optic connections closer to customer premises. This requires massive infrastructure investments, which carriers all over the world are making.

The wireless industry will attempt to capture greater market share by offering new services. Their strategy is to make the cellular handset

---

[43] R. Siklos, "How much profit is lurking in that cellphone?" *The New York Times* (March 5, 2006), BU3.

indispensable as a personal and business tool. One of these services is television.

We are already seeing the start of limited digital TV broadcasts to wireless handsets.[44] Full rollout is being delayed by the usual battles over industry standards, as companies jockey for competitive position.[45] Furthermore, new direct–to–handset TV broadcast technology, which completely bypasses the cellular networks, may limit TV's value to the cellular industry.

Although wireline and wireless are the whole game right now, other ways of reaching consumers are emerging. The Edison-era electrical distribution system could pose a competitive threat to phone companies. The technology for providing broadband Internet access over power lines is improving slowly, offering one more choice among high-capacity "pipes."

New wireless technologies are coming, too. The impending shutdown of analog TV and consequent parceling out of its broadcast spectrum will free up the airwaves for alternatives to cellular. WiMax, a fixed wireless broadband service, is being touted as competition for both cellular and wireline carriers in providing voice and data.

You can be sure that the industry will continue to invest in technology to provide new services. It's just as certain that it will continue to consolidate. Developing technology and building infrastructure is expensive, and bigger providers have more resources to use for these processes.

## *Paying for the network: Tiered service, higher prices*

Whether they operate over wires or through the air, carriers have to streamline and automate their network operations to reduce operating costs. They must also make their networks more flexible to speed up the implementation of promising new services.

This means that migrating to unified IP networks is a priority, although legacy network units will continue to exist alongside them

---

[44] M. Odell, "Mobile television may be the answer," *Financial Times* (February 13, 2006), FTDB1. This service was first introduced in South Korea, an indication of where consumer service innovations are hitting the market first.

[45] J. Yoshida, "Mobile TV may not yet be ready for prime time," *EETimes* (February 13, 2006), 1,13.

for a long time. An IP structure not only delivers flexibility at lower cost, it allows the same network to handle voice and data. New software architectures are now being implemented to achieve these goals.

An IP network offers a strategic advantage to go along with its operational superiority. Executives with the vision to position their companies for the future are already driving in this direction.

Cutting costs and rolling out attractive offerings brings tangible benefits to the top line of the corporate balance sheet. Carriers will also see improved profits if they implement a controversial change in their billing practices: tiered billing, that is, charging different rates for different types of data or different levels of service quality.

In one model, carriers would change their billing to differentiate among classes of data carried on their networks. Voice, for example, might be charged at a higher rate than data.

Another approach under discussion is to build a new, higher-speed network infrastructure, but bill differently for it. "One version would be free, as everything is now. The other would be the digital equivalent of HOV lanes on highways ... to qualify for the speed slot, sites and services would have to pony up huge tolls."[46]

It's only a matter of time before the carriers rationalize their billing to reflect new traffic levels and their associated costs. The demand for bandwidth on public networks is steadily increasing, the carriers are spending huge sums to increase capacity, and tariffs must eventually reflect market realities.

Needless to say, proposals to change the billing model raise violent opposition. The public takes cheap Internet traffic costs for granted, forgetting that the networks are not publicly supported highways. A free consumer VoIP call is not free for the carriers that transport it.

In an unregulated, competitive communications marketplace, however, the tendency will inevitably be for prices to approach marginal costs, generating losses for the least efficient network operators and satisfactory profits for the best ones. The economic challenge is that the marginal cost of transporting an incremental bit is as close to zero as one can get. That's a poor basis for funding the massive investments the carriers need to make to build and maintain their networks.

---

[46] S. Levy, "The technologist: When the Net goes from free to fee," *Newsweek* (February 27, 2006), 14.

Prices for transport must at least cover network operating and capital costs. As long as carriers are privately owned and rely on commercial financial sources, they will have to justify investments on the basis of return on capital. Right now they're justifying their investments on the basis of the returns they expect when the competitive situation reaches equilibrium.

In terms of profitability, content owners still have an early advantage over carriers. The ready availability of broadcast, broadband cable and wireline, and low-cost home digital storage makes access easy to find at very low prices. Valuable content, on the other hand, remains a scarce commodity, and people are willing to pay for it.

Where all this will end up remains an open question. History has shown that sooner or later companies in capital-intensive industries stabilize prices by consolidating facilities and attempting to reduce competition.

But history has also shown that in dynamic industries with extraordinary technological innovations, slowing competition is not an easy option. The communications industry certainly fits that description.

## Digital electronics: Markets, manufacturing, and global strategy

Crankshafts and pistons don't transport us from place to place; roads and cars do. Microchips and lasers don't link people to the online bounty of information and social interaction, or directly to each other. Those are the functions of the Internet, the global communications network, and the electronic devices that connect to them.

The car and the Internet are systems that empower. Efficient transportation and easy access to information and communications have had a profoundly transformative effect on human social and economic life. In the end what matters is how we use our innovations.

Yet without engines and axles there would be no cars. In the same way, without powerful computers in the network, innovative software to manage them, PCs, and fiber optics, there would be no Internet. The manufacturing of electronic products is the foundation of the digital world.

To gain perspective on the scope and significance of the world's electronics industry, we will begin with the smallest components, consider product direction, then look at the changing structure of the industry.

## Semiconductors: A flood of chips and money

We are surrounded by trillions of transistors and billions of lasers, with more coming every year. Semiconductor manufacturers have produced millions of transistors for each person on earth every year since 1985, with each successive year seeing a dramatic increase in the total number of devices in use in electronic equipment.

As shown in Figure 8.4, a commonly quoted statistic pegs the number of transistors in use per person at about 100 million in 2003, and that is expected to reach *ten billion* per person in 2009.[47] By any measure these numbers are staggering. They are proof, if any were needed, of how important semiconductors are to the world economy.

Figure 8.5 bears this out from a different perspective, tracking how much of the value of electronic products has been attributable to semiconductors over the years.[48] While devices accounted for only 4 percent of product content in 1979, by 2006 that figure had risen to about 18 percent. The semiconductor content is closer to 30 percent in products such as cell phones. Manufacturers are turning to smarter, more expensive chips to manage the new features and functions they are adding to their products.

This is especially true in two product categories, which accounted for fully 40 percent of the global sales of semiconductors in 2004 (see Figure 8.6).[49] Wireless products absorbed 22 percent of semiconductor production, while 18 percent went into automobiles.

## New products generate continued growth in electronics

Automobiles are the leading edge of two developing trends: the replacement of mechanical sensors in machinery with solid-state electronic devices; and the control of the key elements of automobiles by onboard computers.

---

[47] Prepared from US Census Bureau data; transistor data from Semiconductor Industry Association (SIA) 2005 Annual Report, p. 14; SIA 2001 data in presentation by Dr. U. Schumacher, ICT Forum Hanover, March 11, 2003.

[48] Chart from T. Thornhill III, CFA, UBS Investment Research (UBS Securities LLC, an affiliate of UBS AG (UBS) – March 2006).

[49] T. Thornhill III, CFA, "Global semiconductor forecast 2006–07," UBS Investment Research, Global Semiconductor (UBS Securities LLC, an affiliate of UBS AG (UBS) – February 1, 2006), p. 9.

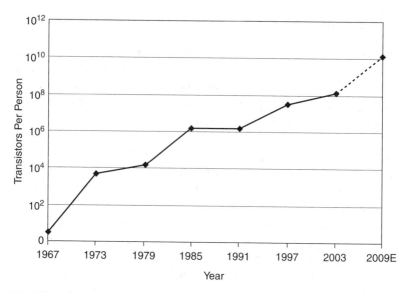

**8.4.** The number of transistors per person on the planet. Data from US Census Bureau and Semiconductor Industry Association (SIA) (ref. 47).

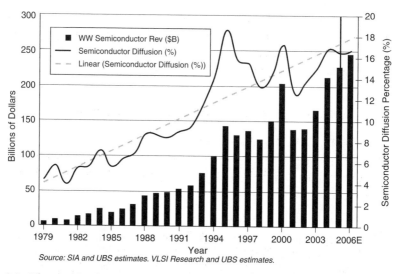

**8.5.** The worldwide (WW) sales of semiconductor devices and the fraction that they represent of electronic systems sales. The dotted line shows the average trend. Copyright © 2006 UBS. All rights reserved. Reprinted with permission from UBS (ref. 48).

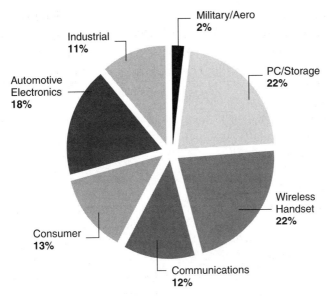

**8.6.** Estimated percentage of semiconductor devices used in major electronic product sectors (2004). Copyright © 2006 UBS. All rights reserved. Reprinted with permission from UBS (ref. 49).

It is estimated that electronic systems account for around $2,500 of the cost of the average automobile, in the range of 10 to 15 per cent of its sale price.[50] These systems range from high-end audio and video entertainment systems to air bag, engine, and antilock brake controls.

More sophisticated control, safety, and entertainment systems are being introduced every year, virtually guaranteeing that the percentage of a car's cost that is attributed to electronics will continue to increase.

The other big user of semiconductors, the wireless industry, churns out new cell phone models on what seems like a weekly basis. Since the mid-1990s, the simple wireless handset has evolved into an all-encompassing appliance combining the functions of telephone and television receiver, Internet and e-mail tool, financial services terminal, still camera, videocam, and fashion statement. And it fits in a shirt pocket.

---

[50] R. DeMeis, "Cars sag under weighty wiring," *EETimes* (October 24, 2005), 47–49.

Informal surveys suggest that most owners see their handset as their primary personal electronic product. The average consumer keeps a handset for about two years. At that point the added features of new models entice the owner into replacing it. As a result, industry sources forecast continued global sales of over 600 million handsets per year, with annual sales figures of a billion units in sight.

Demand for other electronic products also continues to rise, driven by the introduction of new digital functionality and the emergence of completely new technologies. We have digital cameras, personal computers, digital wireless devices, digital television receivers, and digital video and audio recorders.

Analog product categories, even long-established ones, simply disappear when faced with competition from digital devices. Videotape has gone the way of the vinyl LP record. They were both killed by optical disk player/recorders. Digital cameras are rapidly turning the film camera into a museum piece. Film utilization is estimated to be dropping at the rate of 20–30 percent a year.[51]

Even digital devices can be victims of quick obsolescence. Low-price digital cameras, for example, are competing against a formidable foe: the picture phone. Digital cameras built into cell phones now outnumber stand-alone devices.

### Prospects for continued growth

Commercial radio and TV broadcasting are the last major analog holdouts in the consumer market. Digital satellite TV has made significant inroads, and satellite radio has emerged. However, most television shows, even on digital satellite or cable feeds, originate in analog form. On the receiving end only a small percentage of TVs are truly digital. Few households are equipped to receive digital radio by satellite.

That is about to change, with television leading the way. Governments in many countries have mandated a switch from analog to digital television (DTV) broadcasting by 2010.[52] It is virtually certain that TV manufacturers will rack up huge sales of new DTV sets and analog-to-DTV converters for existing sets once the switch-over occurs.

---

[51] "Film fades fast," *Photonics Spectra* (March 2006), 32.
[52] R. M. Rast, "The dawn of digital TV," *IEEE Spectrum* (October 2005), 27–31.

Sales could happen fast, since today's consumers are eager to adopt new technology. Figure 8.1 shows the penetration rates of consumer electronic products in the US since 1947. Color television and PCs both needed about twenty years to achieve a 60 percent penetration rate. It took cell phones about fifteen years to reach the same level of penetration.

DVD players, in the most successful product launch in consumer electronics history, reached the 60 percent mark in only six years. The current pace of penetration suggests that digital television (DTV) will achieve a 60 percent penetration level in under six years.

Terrestrial radio has lagged behind, but digital technology has now been developed for the oldest broadcast service. Digital Audio Broadcasting (DAB), available primarily in Europe, Canada, and Australia, and HD Radio, launched in the US and being tested in Canada and parts of South America and Asia, both promise "CD-quality" sound with less interference. Both require consumers to buy new receivers to get the service.

Given relentless consumer and industry demand for newer, more sophisticated digital technology, it is no surprise that electronic products make up an increasing part of the world's gross national product. As shown in Figure 8.7, this share increased from 0.5% in 1958 to just about 2.3 percent in 2006.[53]

Readers might be surprised at how small this percentage is for an industry that garners so much attention. However, this number covers only the value of the electronic products themselves. If you take into account the huge value produced by business categories that leverage electronic products, such as software, services, and non-electronic equipment, you get a more accurate picture of the industry's importance.

Countries cannot afford to ignore the industry's potential if they wish to sustain or grow their economies. But they must decide how to handle the many challenges the electronics industry presents.

## Deflation in action

Throughout its history, the electronics industry has displayed some consistent if contradictory traits:

---

[53] Data from T. Thornhill III, CFA, UBS Investment. Research (UBS Securities LLC, an affiliate of UBS AG (UBS) – March 2006).

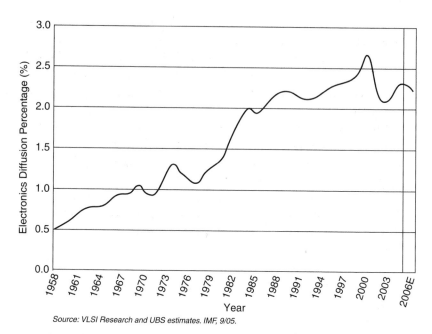

8.7. Electronic products represent an increasing share of the world's GDP. Copyright © 2006 UBS. All rights reserved. Reprinted with permission from UBS (ref. 53).

- steadily increasing product volumes;
- constant advances in product functionality;
- steep declines in selling prices;
- ever-shorter product cycles.

To make matters worse, every emerging technology brings new companies into the industry, heating up an already intense competitive situation. In this environment, sustained profits are elusive for all but the best-managed, most dominant companies in their market sectors.

The standing joke is that the consumer electronics industry loses money on every unit of product it sells, but makes it up on volume. Everyone recognizes the problem, but companies that compete in the global consumer electronics market usually have no choice but to compete on price, or at least meet the lowest price offered.

However, profitability in a new product is generated by being first to market. This puts a big premium on effectively managing innovation to establish market leadership with new products.

Yet even with new products, companies must reduce costs to stay competitive. With Asian companies leading the low-cost parade, competitors in other geographies either follow or drop out. As we have seen, this explains the massive migration of engineering resources and manufacturing from developed economies with a free trade policy to the lowest-cost countries.

If the world appears "flat" to some observers, it is not because the international commercial playing field is level or the game is fair. Governments support their national champions to the detriment of foreign competitors. How can a company already running on thin margins counter that kind of pressure?

One answer: electronic product companies with global ambitions must be large enough to support product diversity, scaled R&D facilities, and the lowest-cost manufacturing in their sectors. That's why markets such as TV receivers, memory products, and network equipment, where industry standards determine many product features, have seen especially rapid consolidation through mergers or the disappearance of participating companies.

In the network equipment space, for example, Cisco Systems has become the dominating force in the world market as competitors have dropped out or been relegated to niches. Samsung is now the leading manufacturer of FLASH memories, as Seagate is of magnetic disks, and Intel is of high-end microprocessors.

All of these companies can use their size and economic strength to create the high-quality, low-cost products that boost sales volumes. With the demand for consumer electronics opening new markets around the world, it would seem that their opportunities for international sales are enormous.

And so they are, if the companies meet a fundamental challenge. Rising incomes in the developing economies mean that billions of people are entering the market for electronic products – if they are cheap enough.

While we are used to $200 cellular handsets, a $40 handset is what a Chinese worker can afford, and a PC for $100 would be a wonderful buy. These end-user prices can be met profitably only by product vendors who understand how to design with minimal features and low cost in mind right from the start.

It's not just consumer products that present a challenge. In emerging markets even infrastructure products are under heavy cost pressure.

Consider the economics of selling telecommunications equipment in India, for example. Basic cellular phone service in India is offered very profitably at only about $10 a month, compared to about $40 in the US. Such low-cost service is possible for two reasons.

- Employees of the carriers in India get about one-fifth the wages of their US-based counterparts.
- Indian carriers buy their equipment and associated software from international vendors under very favorable terms. The vendors have to meet very low price points to win the business.

The chief financial officer of Siemens Communications, an equipment vendor to the wireless industry, gave the best summary of the situation: "There is no such thing as easy mobile money anymore."[54] Just as in the wireline and cable industries, any investment in new equipment to increase the capacity of a wireless network must be justified by new revenues. As these are hard to predict, the pressure is on equipment vendors to provide products at the lowest possible cost.

Equipment manufacturers thus have a choice – forego big sales opportunities in the fastest growing markets, or stay in the game by reducing all enterprise costs to preserve profits while accelerating the speed with which new products are brought to market.

Executing this strategy is not simple. To summarize its elements, as covered in Chapters 4 through 6, companies need:

- A global presence to help find the best ideas, technical talent, and manufacturing facilities.
- Fast product cycles, critical to sustaining sales momentum in a dynamic market.
- High-quality engineering for efficient R&D.
- Lowest labor costs consistent with tasks to be performed.

As we concluded in Chapter 6, there is no single right approach to locating engineering resources or setting manufacturing strategies. However, protection of intellectual property and the shortening of supply chains to speed up product cycles can easily outweigh other factors in selecting locations.

One thing is certain: there are more potential manufacturing locations for technology companies than there were twenty years ago.

---

[54] Quoted in J. Yoshida, "Mobile operators hammer on costs," *EETimes* (February 20, 2006), 1.

Where they choose to locate will shape the course of the world economy for years to come.

In the next (and final) chapter we will offer some thoughts on how those choices are being or should be made, and discuss their implications for developed and developing economies alike.

## Evolving into the future

If the electronics industry (including software and hardware), communications, and the Internet seem to be generating turmoil and progress in almost equal measure, that's not surprising. They are progenitors of and participants in a revolution. Turmoil goes with the revolutionary territory, and progress is the outcome of a good revolution.

It's fascinating to see how interdependent these three areas are. Advances in the electronics industry enable new products and services. These are adopted by the communications industry, which makes the new services available to subscribers. The Internet helps inform the public about the products and services, creating demand and hence huge markets.

They participate in an accelerating yet circular process. Electronic innovation continues to drive down the cost of products and services. The public becomes comfortable with these more quickly with every new round of introductions, thanks to their price and their improved ease of use. As a result, users are more willing to buy the next round of changes or improvements, further shortening the time to mass market adoption. New services, in turn, drive up demand for communications and storage capacities, which increases the impetus for more electronic innovation.

Dizzying as it has been, the recent history of electronics, communications, and software has taught us one thing: with obsolescence a way of life, only the most innovative companies can thrive. There is no room for the slow or the risk-averse.

That applies to nations as well, as we will see in the following pages.

# 9 | *The digital world: A global village*

MY career as a researcher gave me a front-row seat during the period when electronics quickly advanced from primitive prototype transistors and lasers to incredibly tiny devices, sophisticated software-driven systems, and vast computing and communication networks. It was an exciting time to be a scientist, and to make my own contributions to the progress of the field.

As a venture capital manager, I've had an equally fortunate vantage point. I saw digital technology become a global force, delivering new services and creating opportunities where none had existed before. I also had the opportunity to arrange funding for innovative companies that helped move the technology forward.

No one in the 1950s and 1960s, when the fundamental discoveries were rolling out of US corporate laboratories, could have imagined that digital electronic technology would help to lift whole countries from isolation and poverty into participation in the global economy. But I've seen first hand how these revolutionary innovations have improved the human condition and even realigned the economic order among nations.

Today we live in a global village sustained by digital computing and communication technologies. It is a much more complex world, technologically, socially, and politically. To conclude this study of the nature and influence of digital technology, we will look at what the globalization of technology means for developed and developing nations alike, and how innovation and manufacturing can build prosperity at both stages of economic progress.

We'll begin with an assessment of the current situation.

## Trade realities

Free trade has served the world well. Under its aegis, digital technology has spread around the world. It has not only benefited hundreds of

332

millions of users, but it has also created jobs and industries for workers and engineers in newly emergent economies. These workers, in turn, produce the low-cost electronics that are used by people across the globe.

For many, however, free trade and the globalization of electronics has brought new problems. Workers in developed countries who lose well-paid production jobs to low-cost foreign labor may face a future that is far from rosy. There is growing concern in the developed world that the exporting of jobs is hurting national prosperity.

Reality, as usual, is more complicated than the perceptions. Figure 9.1 gives protectionists and free trade proponents alike something to think about. It shows the changes in relative market shares among major economic powers for high-tech products between 1980 and 2003.[1] It also lists total trade figures for each of those years.

No one who has seen the statistics in this book and elsewhere will be shocked to see that the emerging Asian economies (China, South Korea, and others in the region) have substantially boosted their share of the world's technology product exports, from 7 per cent in 1980 to 28 per cent in 2003. Those gains had to come from somewhere. During the same period, the US share declined from 30 to 16 per cent, and that of the EU from 41 to 32 per cent.

In terms of relative market share, then, the developed world has lost significant ground to the emerging nations. But market share is not the whole story. Total dollar figures give us an altered perspective.

World trade in high-tech products totaled $190 billion in 1980. The figure for 2003 was ten times higher, at $1,902 billion. (All figures are in 1997 dollars.) While the US share got smaller in percentage terms, the monetary value of its share has risen from $60 billion to $200 billion.

In other words, so long as the total market is expanding, this does not have to be a zero-sum game. You can do quite well with a smaller slice of a much bigger pie. But you don't want your share to decline indefinitely. It is important to spot trends that could pose a threat to future prosperity.

---

[1] From "The knowledge economy: Is the United States losing its competitive edge?" *Benchmarks of our innovation future; the report of the task force on the future of American innovation* (February 16, 2005), p. 11 (accessed at www.futureofinnovation.org/PDF/Benchmarks.pdf on May 9, 2006).

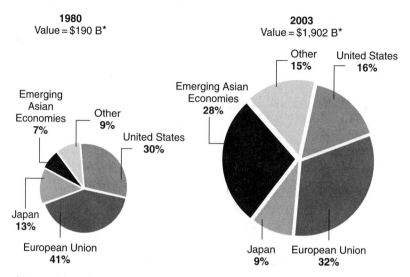

* 1997 US dollars. **Emerging Asian Economies:** China, South Korea, Taiwan, Singapore, Hong Kong, India. **High-tech** includes Aircraft, Pharmaceuticals, Office and computing machinery, Communication equipment, Medical, precision, and optical instruments.

Source: National Science Foundation, Science and Engineering Indicators 2006, Appendix Table 6-4. Compiled by the Association of American Universities and APS Physics (Washington Office).

**9.1.** Fraction of world trade in high-tech products in 1980 and 2003 by region of origin (from www.futureofinnovation.org, "The knowledge economy," ref. 1. Updates of the 2001 data to 2003 have been provided by the American Physical Society (APS)).

## Asian tigers: How much of a threat?

One of those trends, of course, is the migration of manufacturing to Asia. This has dramatically changed world trade.

Manufacturing is fueling the rapid industrialization of the leading Asian economies, as we discussed in Chapter 6. Their success is no accident. Asia has driven the process of growing high-tech industries through national programs to expand education, fund national R&D programs, build national laboratories, and subsidize industrial projects.

Of course, programs like these can be found in developed countries as well. In all cases the results are largely dependent on the quality of their management in producing economic value, and the competitive advantages of the affected industries.

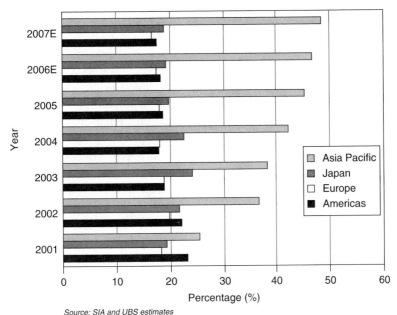

Source: SIA and UBS estimates

**9.2.** Semiconductor device sales by region as a percentage of total worldwide semiconductor sales since 2001. Copyright © 2006 UBS. All rights reserved. Reprinted with permission from UBS (ref. 2).

Another trend that indicates the success of Asian countries in creating an electronics manufacturing industry is their rising consumption of semiconductor products. Figure 9.2 shows that about 45 percent of the devices produced in the world ended up in the Asia Pacific region in 2005, almost double the 24 percent of production that shipped there in 2001.[2] This shows that the developing economies of Asia are producing more of the world's electronic products, of which semiconductors are primary components.

### Reverse growth

The shift of domestic manufacturing to offshore locations is not a new trend. The assembly of electronic products like TV receivers first

---

[2] T. Thornhill III, CFA, "Global semiconductor forecast 2006–07," UBS Investment Research, Global Semiconductor (UBS Securities LLC, an affiliate of UBS AG (UBS) – February 1, 2006), p. 10.

migrated out of the US in the 1970s – to Mexico. But the floodgates really opened in the 1990s, when China became a source of low-cost manufacturing.

The decline in US manufacturing activities has now become painfully obvious. Since 2000, more factories have closed in the US than have opened. Investment figures show a declining amount going into new plant construction. Only $18.7 billion was invested in plant construction in 2005, compared to $43.7 billion in 1998.[3]

Just as troubling, the decline in manufacturing investment has spread to other industry sectors. Process industries are following high-tech ones overseas. The US chemical industry, with annual revenues of $500 billion, is one example. Only one new billion-dollar chemical plant was under construction in 2005, compared to fifty such plants in China.[4]

But it's not just the US and other established, developed economies that are affected by the wholesale migration of manufacturing to China. Even Brazil is feeling the effects, as Chinese goods pour into their market.[5]

### Difficult balance

The growing power of Asia relative to the developed world is also reflected in the shifts in the balance of trade among the countries of the world. As Figure 9.3 makes clear, the US trade balance for high-tech products turned negative in 2001.[6] It has been in decline ever since. The deficit with China alone reached $200 billion for 2005–2006 (including all products).

While Japan and Germany with strong manufacturing sectors continue to have trade surpluses, other major countries such as the UK, France, and Spain also have significant trade deficits with Asia.[7]

---

[3] T. Aeppel, "US 'birthrate' for new factories is steadily falling: Decline prompts concern about long-term health of nation's manufacturing," *The Wall Street Journal* (March 13, 2006), A4.

[4] "No longer the lab of the world: US chemical plants are closing in droves as production heads abroad," *BusinessWeek* (May 2, 2005), 80.

[5] D. J. Lynch, "China's growing pull puts Brazil in a bind: South American nation swamped by imports," *USA Today* (March 21, 2006), 1B, 2B.

[6] From www.futureofinnovation.org, "The knowledge economy," p. 11.

[7] F. Norris, "Setting a dubious record, but still out of line," *The New York Times* (March 18, 2006), C3.

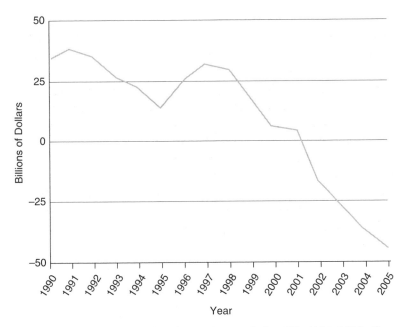

**9.3.** Trade balance in high-tech products of the US 1990–2005 (from www.futureofinnovation.org, "The knowledge economy," ref. 6. Updates of the 2001 data to 2005 have been provided by the American Physical Society (APS)).

In one sense the US is fortunate, because at least its huge trade deficits are financed by the importation of capital. There are even those who argue that the deficits are irrelevant to the future of the US economy.

More sober analysts point out that no trade deficit of this size is sustainable indefinitely, especially when it goes largely to support consumption rather than capital investment. It is widely expected that the value of the dollar relative to other major currencies will depreciate if this deficit continues unchecked.

Experts also predict that, as the results of the deficits become apparent to the public, there will be mounting political pressure to "do something." The public in the US and other affected countries already fear that their economies are at risk of reverting to the Stone Age as their advanced industries desert them for low-cost Asian sites.

Given those depressing statistics and dire forecasts, their anxiety is understandable. Whether it is justified is another matter. The problem is that we don't have adequate information on the true deficit, or the ability to interpret it.

We've already seen that while part of the migration of manufacturing to Asia goes to the region's native industries, a substantial percentage of the facilities there are owned by international technology companies. These companies have moved operations overseas to get access to the low costs and available talent that keep them competitive in the world market. The trade deficit figures do not account for their production.

For example, US corporations with production sites in Asia will import many of their own-branded products back into the US. The more finished products the company re-imports and sells domestically, the higher the trade deficit.

To further complicate the situation, US-branded products that are both manufactured and sold overseas never show up as a US export, even though they obviously benefit the companies that make them.

Analyst Marc Chandler, a currency strategist at Brown Harriman in New York, believes that half of the trade deficit reflects overseas transactions among subsidiaries of US-based companies.[8] Hence the trade deficit number, while real, is misleading. It masks the movement of products within US-headquartered companies. We must take care in judging the country's economic condition based on the official deficit figures.

## Global situation, global solutions

We come back finally to Joseph Chamberlain's lament of 1905, quoted in connection with manufacturing – how does an advanced economy with relatively high labor costs generate new industries fast enough to replace those that migrate into lower-cost countries?

Not to worry, responds US Commerce Secretary Carlos Gutierrez. He argues that the US, for one, has nothing to fear from the rise of China as a low-cost manufacturing power. Instead, the country should concentrate on further growth in high-tech, high-value industries.[9]

---

[8] Quoted in C. Karmin, "Theories on what bedevils beleaguered dollar," *The New York Times* (April 3, 2006), C7.

[9] Quoted in "China output not a threat, official says: US urged to focus on its strengths," *The New York Times* (April 1, 2006), C4.

Secretary Gutierrez is stating a belief commonly held in advanced economies: new high-tech, high-wage industries will spring up to replace older ones being cannibalized by offshore competition.

They don't spring up on their own, of course. They need encouragement and resources. While some look to the government for subsidies and technology development programs, many others, presumably including Gutierrez, believe "lower taxes and more business investment can do a better job delivering the competitiveness America needs."[10]

Economic analyst R. J. Samuelson, a contributing editor to *Newsweek* and regular columnist for *The Washington Post*, takes a more philosophical approach. Samuelson maintains that America's ability to compete is based on cultural advantages.

America's economic strengths lie in qualities that are hard to distill into simple statistics or trends. We've maintained beliefs and practices that compensate for our weaknesses, including ambitiousness; openness to change (even unpleasant change); competition; hard work; and a willingness to take and reward risks. If we lose this magic combination, it won't be China's fault.[11]

Gutierrez and Samuelson both have a point. Innovation, as we've seen throughout this book, is a proven path to economic prosperity. The innovative strength of the US has served the country well in the past, and should continue to do so. As the country that invests the most money in R&D by far, and essentially created the venture capital industry that funded many of the innovative new companies of the digital revolution, we have a tradition of developing the next big, commercially successful technology.

The question is whether the past can still be an adequate predictor of how fast major new industry formation will occur in developed countries like the US. These nations need sources of high-value exportable products to replace departing industries that once employed hundreds of thousands of workers.

Innovations are counted on as the engines for growth. But, even if innovations appear, there's no guarantee they will originate in the US or anywhere else in the developed world. We now face competition

[10] T. G. Donlan, "Technological hubris," *Barron's* (June 19, 2006), 51.

[11] R. J. Samuelson, "Sputnik scare, updated," *The Washington Post* (May 26, 2005), A27.

from countries that have worked hard to develop their own innovative capabilities. We are no longer alone in the sphere of innovation.

## Assessing the environment

To plan for the future, we in the developed world must make a proper assessment of where we are now, and how we got here. We can start by accepting the fact that the worldwide diffusion of R&D and high-tech manufacturing has created a new set of competitive realities. These can be summarized in seven points, all covered at length earlier in this book.

- History has shown that countries that begin with low-cost manufacturing strive to move up the value chain over time.
- Competition for the creation of high-value-added industries is now global.
- Government subsidies (hidden or open) are prevalent everywhere to encourage high-tech industrial development. Their effectiveness, however, varies greatly.
- At the corporate and national level, the challenge is to exploit innovations *rapidly* to create maximum value. Time to market is a key competitive differentiator.
- Fierce international competition in commercializing innovations will be a way of life.
- Developing valuable intellectual property is crucial. It, too, is now a global process. It is no longer the exclusive birthright of the developed economies. This makes its protection far more difficult. Hence the necessity of rapidly commercializing innovations, to keep in front of the pack.
- Manufacturing must be considered as important a source of value creation as the development of intellectual property. Industries are most valuable, and sustainable, when they implement a large part of the value creation process, from innovation to product delivery. Without participation in those important aspects of manufacturing that leverage sophisticated intellectual property, the value extracted from certain innovations is limited and may be short-lived.

The alert reader will notice that the last two points on innovation and manufacturing are the most detailed, and the last point the most prescriptive. In my opinion those two activities, and the tension between them, supply the strategic basis for continuing prosperity.

Before we discuss these final points at length, however, let us go through the earlier ones in order. We can dispatch the first three quickly enough.

By now we know that Korea, Taiwan, India, and China are either making the move, or have moved up the value scale from low-cost manufacturing to high-tech electronics. Korea is even challenging Japan, which years earlier had displaced the US as the major manufacturer of consumer electronics.

China's current dominance at the bottom of the value chain makes it difficult for other countries to make inroads there, but this will change as China leaves low-skill assembly behind in favor of higher-value industries.

A more crucial question is how to extract maximum value from industrial innovations. For developed economies, competitive advantage comes from technological superiority, competitive cost, and the development of unique products serving large international markets, plus the ability to quickly exploit innovations in the marketplace.

As the products mature and competition emerges, developed countries decrease costs by automating manufacturing processes. Eventually, when skilled labor is no longer required for production, the products migrate to low-cost manufacturing nations.

This seems like a reasonable approach, especially since, sooner or later, high-tech industries in most Western countries have to fend for themselves. In theory, they will have embarked on the commercialization of new, higher-value innovations by the time older products migrate to low-cost manufacturing sites.

## Policy and politics

We now turn to the issue of productivity. How do companies in the developed economies achieve their objectives in the face of efficient new competitors? This is a much-debated question. The standard answer in the US and other developed economies is to:

- increase the technical capabilities of the workforce through training and education;
- build new industries on the basis of advanced R&D, which is heavily supported by taxpayers.

People who support these recommendations reflect the mindset that R&D will generate new industries, that these industries will not be subject to limitations on labor costs, and that the workforce will be better educated in math and the sciences than the current norm. They also assume that advanced intellectual property can be protected from piracy as it generates valuable new large industries.[12]

This way of thinking commonly ignores the need for domestic manufacturing in critical areas of innovation. Thus it implies the abandonment of all but the most esoteric manufacturing activities. Non-esoteric manufacturing would be outsourced. Most industries would be built around exportable services and marketable intellectual property.

There are unfortunate social ramifications in such a system, especially for employment patterns. No one denies that highly-skilled employees are essential to advanced product and service industries and would be well paid by international standards. But what would happen to the rest of the workforce?

Moderately skilled workers might get jobs in service industries at much lower salary levels. Faring worst in this scenario are low and semi-skilled workers, who will wind up competing for work with people in low-wage countries.

Another problem is the emphasis on R&D funding. Proposals for improving the competitive position of the US in high-tech industries commonly include recommendations for increased R&D spending in the physical sciences.[13]

But as we have documented, current federal R&D spending is heavily weighted toward supporting the life sciences and military projects. As there is no guarantee that this mix will change, new initiatives will require large increases in funding.

Finally, these proposals also neglect what I consider to be the most crucial part of the innovation process. In order to meet international

---

[12]  A good example can be found in the proposal from the Committee on Prospering in the Global Economy of the 21st Century, *Rising above the gathering storm: Energizing and employing America for brighter economic future* (Washington, DC: National Academy of Sciences, National Academy of Engineering, and Institute of Medicine, 2005).

[13]  Examples include the report prepared by Committee to Assess the Capacity of the US Engineering Research Enterprise, *Engineering research and America's future: Meeting the challenges of a global economy* (Washington: National Academies Press, 2005); and the report of the Task Force on the future of American innovation, "The knowledge economy: Is the United States losing its competitive edge?" (accessed at www.futureofinnovation.org, February 16, 2006).

competition in exploiting the commercial value of many innovations, developed economies like the US must know how to make what they invent.

Markets are created by selling innovative products and adding value at various stages of the delivery chain. Manufacturing creates a lot of the value in such product categories as advanced semiconductor production. In these fields the companies that command the most advanced processes are in the best position to quickly enter the market with the highest value products.

In short, the innovation process is crippled without at least some critical manufacturing activity. Innovative physical products (as opposed to software or media content) only generate profitable mass-market launches if part of the product value is contributed by innovative manufacturing.

## Strategies for sustaining progress

All of these considerations focus us at last on the most important of the competitive realities listed above. They address the two areas where economic prosperity is created: developing valuable IP, and manufacturing products to exploit it. Countries that encourage these activities stand the best chance of maintaining or increasing their industrial strength even in the face of challenges from emerging nations.

Many of my remarks here will refer to the US and other developed countries. However, they could just as easily apply to any country seeking sustained economic growth and prosperity. I do not pretend to have all the answers, but all of the facts marshaled for this book support these conclusions.

### *Innovation: R&D teams with manufacturing*

The US economy is huge, so it needs hugely successful products to compete in global high-tech markets. R&D is a critical step in developing such high-value products, but it must be directed at areas of great economic promise.

There is no lack of candidates. This book has focused on electronics because they have long proven their importance in global commerce. Software will continue to be an important value generator, but other technologies with vast potential future value are now being developed, all based on the physical sciences.

For example, as scientists uncover new information on near-atomic level phenomena, a field broadly called "nanotechnology" has emerged. It holds the promise of developing several new industries, including new kinds of electronic devices, advanced structural materials, special-purpose coatings, and biomedical molecules.

Genomics is another burgeoning field of exploration. The field is narrower, but its implications are profound. It seems certain to create whole new modalities for disease control and human development.

Perhaps the most urgent and exciting area of exploration is renewable energy. One of the most promising sources of energy is the solar cell, an energy converter that produces electricity from sunlight. This device is fertile ground for a national R&D effort capable of generating new domestic industries, and it exemplifies the need for manufacturing as a national priority.

Solar cells are very large semiconductor p-n junctions, a technology we understand well. It so happens that this technology is highly material science-intensive, requiring massive investments in production plants resembling larger versions of those used for chip production.

Research to improve cell efficiency is already being done in the US using federal funding. However, it will lead to measurable market value only if production plants are also built domestically. Otherwise these innovations will be marketed by those that have made manufacturing investments in Asia or elsewhere. The US R&D investment will have a negligible effect on exports or balance of trade.

That is the crux of the matter. Building large industries based on the physical sciences requires manufacturing commercial products at high volumes and acceptable costs. These industries develop by the trial and error process that only comes from a close link between manufacturing and R&D. The countries that benefit the most from materials and process-intensive innovations are those that turn manufacturing into a competitive advantage.

As discussed in Chapter 6, the semiconductor industry is a good model for how this works. Its process-intensive nature has kept it in close touch with the factory floor. Over the years the expertise gained through this tight integration of R&D with production has allowed the rapid introduction of evolutionary innovations into commercial products.

If more proof were needed, the Asian countries now expanding their economies provide it. They have focused on manufacturing as a discipline. This didn't just make them better manufacturers; it's making them innovators, and reducing their dependence on the foreign IP they have been using to build their industries. The IP they have developed on their own is manufacturing-focused, as it was in the US at the beginning of the semiconductor era.

Another excellent example is the flat-panel display industry. While the technology was originally pioneered in the US and benefited from large government-supported programs before 2000, the industry is no longer in its place of origin. Asian countries have poured many years of production innovations into flat-panel devices, ensuring that not a single flat-panel TV or computer display will come from any place but Asia. No one else has the process technology to achieve the necessary yields.

The US, and indeed all developed economies, must maintain an edge in the cycle of product innovation. Small programs to promote manufacturing innovation in the US have been funded through the National Institute of Standards and Technology over the years. But these programs are too small to have a significant impact on US industry.

I am not proposing that we subsidize obsolete plants or inefficient manufacturing. This is bound to fail. Low-wage countries can risk a little inefficiency in their manufacturing process. Developed economies can't. We have to rethink every aspect of the production process to achieve efficient manufacturing through flexibility, improved quality control, and the elimination of routine hand assembly. Those industries that fall behind in manufacturing innovation become easy targets for new entrants.

## Workforce considerations

Sophisticated manufacturing requires an equally sophisticated workforce. To create such a labor force we must have better training. This does not mean drilling line workers in physics and trigonometry. It means developing highly-skilled workers who are comfortable in the most modern production environments.

Inadequate worker skills are not just a production problem. They are an economic issue as well. If the US does make a serious effort to rebuild its manufacturing base – and I believe it must, if it is to save its innovative capabilities – it must face the question of labor costs. There are no easy answers. If labor costs make up a substantial portion of the cost of a particular class of product, then this country cannot compete in that market, because our labor is simply too expensive.

That essentially eliminates opportunities for semi-skilled workers in manufacturing, at least in developed countries like the US. These workers will be displaced by more efficient automation to keep labor content down. The tin-banger in an auto plant is out of a job. The robot operator may not be. Automated plants in developed countries can compete; semi-skilled factories cannot.

## Some concluding thoughts

New nations are emerging as global economic powers, while established economies search for ways to sustain their standard of living. Digital technology is helping to level the playing field. Governments espouse free trade, but try to protect their domestic industries, with varying degrees of success.

Globalization cannot be stopped. The best way to maintain prosperity and growth is to sustain the capability to innovate in all areas, including software as well as advanced manufacturing.

Promising new technologies loom on the horizon. They offer the hope of great new industrial growth, continued improvement in the human condition, and the freeing of still more nations from the cycle of isolation, poverty, and ignorance. We cannot know which, if any of them, will replicate the impact of digital electronics.

We need to give researchers the freedom to pursue innovation opportunities in these fields, rather than being shackled to making incremental improvements for current products. Freeing the creative innovator to explore ideas is a prerequisite for success. When a researcher does come up with a promising commercial innovation, we must make sure that its development is tightly integrated with manufacturing.

The US launched the digital age through a potent combination of innovation, risk capital, and advanced manufacturing. Creating an industry with global ambitions, one that breaks technological barriers and expands economic opportunity, requires all three of these elements. That's the ultimate lesson of the digital revolution.

We have come a long way from that magic moment when the first transistor actually worked. Today's digital technology is influential and successful because it is both powerful and flexible. We should strive for the same qualities in industrial policies.

# Appendix 1.1: Smaller, faster, more efficient MOSFETs

## Bipolar beginnings

The bipolar transistor was the first important commercial device of the 1950s. Figure A-1.1.1 is a schematic of an n-p-n bipolar transistor. It consists of two p-n junctions back to back. The two n-type regions, called the *collector* and the *emitter*, are separated by a very thin p-type semiconductor layer, referred to as the *base*.

This structure is the successor to the original point-contact transistor invented at Bell Labs. The p-n junctions make it far more robust and practical than that first pioneering device.

We can get an appreciation of the bipolar transistor's principles of operation by applying a negative voltage to the n-type top side (emitter region) and a positive voltage to the n-type bottom side of the structure (collector region), where we see a resistor in series with the battery.

Because of the negative (forward) bias on the emitter-to-base p-n junction, electrons are injected into the base region. Since the base is very thin (a fraction of a micron), the injected electrons traverse it with little loss. They are collected by the reverse-biased base-to-collector p-n junction.

The result is that a large current flows through the collector circuit with the large resistor. On the other hand, the current through the other circuit (the emitter-to-base circuit) is very small, because few electrons flow through that circuit. In effect we have built an *amplifier*, in which a small current in the emitter-to-base circuit is translated into a large current in the collector-to-base circuit.

The transistor can also function as a *switch* – we can turn the current flow in the output of the device on and off by turning the emitter voltage on and off.

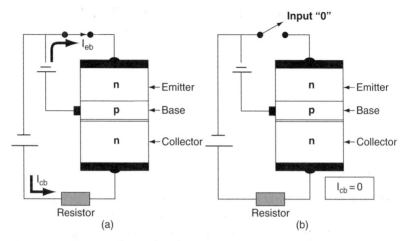

**A-1.1.1.** Operation of n-p-n bipolar transistor. (a) Current flow as emitter-to-base p-n junction is forward-biased. The base to collector p-n junction is reverse-biased. (b) Opening and closing switch turns transistor on and off.

### Advancing to MOSFETs

The second major type of transistor is the MOSFET (metal oxide semiconductor field-effect transistor).[1] This device is the foundation of most of today's integrated circuits, memories, commercial transistors, and imaging devices.

The MOSFET performs switching and amplifying functions similar to those of the bipolar transistor. It is easier to manufacture, and dissipates far less power when it acts as a switch.

In the interest of easy understanding we will focus on the switching function of the MOSFET. That also happens to be the feature of most importance in digital systems.

Hundreds of millions of MOSFETs, miniaturized to the point that their sizes approach atomic dimensions, can be built and interconnected to fit on an integrated circuit chip the size of a fingernail. Such microprocessors are at the heart of today's computers.

---

[1]  S. R. Hofstein and F. P. Heiman, "The silicon insulated-gate field effect transistor," *Proceedings of the IEEE* 51 (1963), 1190–1202. Also see D. Kahng, "A historical perspective on the development of MOS transistors and related devices," *IEEE Transactions on Electron Devices* ED-23 (1976), 655–657.

Figure A-1.1.2 shows a *highly* simplified schematic useful for understanding the basic operations of MOSFETs.[2] The body of the device shown in the illustration (called an n-channel transistor) consists of a p-type semiconductor in which two p-n junctions have been formed by the addition of n-type regions. These are separated by the gate width (distance L). Note that the gate is covered with an insulating *silicon dioxide* (glass) layer overlaid with a *conducting* contact region.

When the gate voltage is zero (a), the source and drain are isolated and the transistor is switched off. To turn the device on, we apply a positive voltage (10 volts in the examples we use) to the gate (b). This produces a *temporary* conducting channel between source and drain, because free electrons are attracted into the gate region that separates them.

Hence we have a switching device where the current between the source and drain is controlled by the value of the applied gate voltage. This functionality comes at a price: switching the device on and off dissipates power, which translates into more heat as we increase the switching frequency. Heat must somehow be removed from the chip in order to keep its temperature under control.

By way of compensation, however, the smaller we make the transistor, the less power it dissipates individually for a given switching frequency. Since the object of microelectronics is to make the transistor as small as possible, switch it as rapidly as possible, and dissipate minimal total chip power, this is a real benefit. It opens the way to putting more and more fast transistors on every chip.

## Shrinking MOSFETs

As noted in Chapter 1, shrinking MOSFETs not only makes it possible to put more of them on a chip, but also greatly improves their performance. To understand the basis of scaling, we use Figure A-1.1.3.[3]

This figure shows two versions of the same transistor, the original device on the left and a scaled-down version on the right.

---

[2] G. McWhorter, "How digital circuits make decisions," *Understanding digital electronics* (a Texas Instruments Learning Center Publication, 1978), pp. 2-12–2-13.

[3] Y. Taur and T. H. Ning, *Fundamentals of modern VLSI devices* (Cambridge UK: Cambridge University Press, 1998), p. 165.

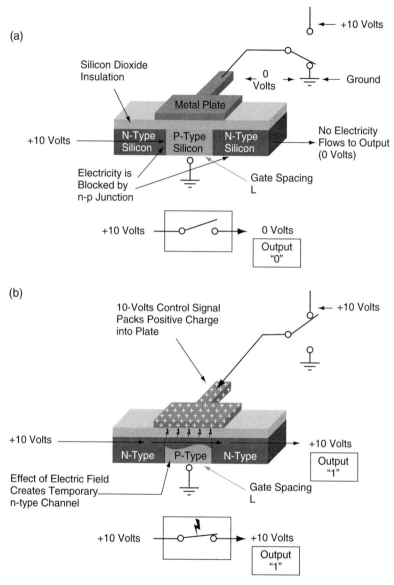

**A-1.1.2.** Schematic of elementary n-channel MOSFET (metal oxide semiconductor field-effect transistor). (a) Gate voltage is 0 and device is off; (b) 10 volts applied to gate turns the device on. Courtesy of Texas Instruments (ref. 2).

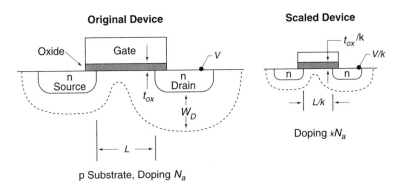

**A-1.1.3.** Principles of MOSFET scaling where the gate length L is reduced to L/k. H. Dennard, F. H. Gaensslen, H. N. Yu, V. L. Rideout, E. Bassous, and A. R. LeBlanc, "Design of ion-implanted MOSFETs with very small physical dimensions," *IEEE J. Solid-State Circuits*, SC-9, p. 256. Copyright © 1974 IEEE (ref. 3).

In the scaled-down version, the linear dimension L, separating source and drain is reduced by a factor of k. The principle of scaling MOSFETs consists of reducing the device operating voltage V and the device horizontal dimensions by the same factor k, as well as the thickness $t_{ox}$ of the glass insulator sitting on top of the gate.[4] In addition, the silicon dopant concentration $N_a$ is increased by k.

Three benefits derive from scaling down.

- Faster switching speed, allowing higher operating frequencies by a factor of approximately k.
- Reduced transistor area by a factor of $k^2$.
- For a single on-off cycle, reduced *switching* power dissipation per transistor by a factor of $k^2$.

How does scaling reduce the power dissipated in switching? The total power dissipated in operation is given by the product of the switching frequency, applied voltage, and the current flowing between the source and drain when the transistor is in the conducting (on) condition. Since the voltage is reduced by k, as is the current flowing between source and drain (because of the smaller device area), the power dissipated is reduced by a factor of $k^2$ for a single switching cycle.

---

[4] Taur and Ning, *VLSI devices*, pp. 164–173.

An example will show how scaling works. Assume that k is 2. The area of the transistor is reduced by a factor of 4; hence the transistor density on the chip can be increased by approximately that amount. At the same time, the switching speed is approximately doubled while the total power dissipated by the chip remains constant. This is basically Moore's Law in practice. It is obvious why shrinking transistors have been the productivity driver of the chip industry.

However, as noted in Chapters 1 and 2, there are limits to the ability to reduce the gate spacing L, set by the fact the two p-n junctions do not stay electrically isolated as the gate length shrinks substantially below 0.1 micron (100 nm). As a result, a leakage current between source and drain exists even when the transistor is supposed to be in the off state, causing power dissipation problems.

A further factor to keep in mind in limiting the *ultimate practical* switching frequency achievable with a MOSFET chip architecture is that increasing the switching frequency increases the total chip power dissipation, and can lead to unacceptable chip heating. Currently, limitations for the smallest, fastest devices are in the 3 to 4 GHz range.

Over time, the complexity of the devices has kept increasing, with more than twenty separate layers of material that have to be deposited and patterned in order to produce a state-of-the-art MOSFET structure and provide the massive on-chip interconnections.

Figure 1.3 shows the cross-section of such a transistor. To get an idea of the scale, note the gate dimension L, which is now below 0.09 microns (90 nm). Extraordinary precision in manufacturing is needed as patterns are repeated millions of times on a chip.[5]

## Limits to progress: Leakage current

Chapter 1 points out that when we reach near-atomic spacing, we approach the limits of our ability to shrink MOSFETs as currently understood. As Figure 1.4 shows, the steady reduction in gate length is putting us close to the mark.[6] It was 0.5 micron (500 nm) in 1993, and is projected to be only 0.032 micron (32 nm) in 2009.

[5] J. R. Brews, "The submicron MOSFET," in S. M. Sze (ed.), *High-speed semiconductor devices* (New York: A. Wiley-Interscience Publications, 1990), p. 144.
[6] T. Thornhill III, CFA, "Global semiconductor primer," UBS Investment Research, Global Semiconductor (UBS Securities LLC, an affiliate of UBS AG (UBS) – March 30, 2006), p. 82.

Let's look at Figure A-1.1.3 again to discover the reason for this difficulty. When we reduce the distance L between the drain and source to below 0.1 micron (100 nm), we start to get significant electronic interactions between them. As we go deeper into submicron dimensions, the atomic interactions increase, making them less and less isolated *electrically*, even when the applied gate voltage is zero.

This results in a small but constant flow of electricity through the device. Called *leakage current*, it flows even when the device is supposed to be in its non-conducting "off" mode. As leakage current flows through the resistive regions of the device, it produces power dissipation (given by the square of the current multiplied by the resistance) which can raise the temperature of the chip to unacceptable levels even in the non-switching condition. Remember that we have many millions of such transistors closely packed together on that chip, most of which are not in operation all the time.

Essentially, the more gate lengths go below that 0.1 micron (100 nm) mark, the higher the static leakage current, power dissipation, and heat they produce. We have encountered a limit to continued progress. The question is, how fundamental is this limit?

One should never underestimate human ingenuity. Several schemes are being used to push device leakage lower and lower. Very sophisticated processing techniques exist to control the doping at almost atomic levels, and novel insulators are replacing the well-established silicon dioxide structures.

**Limits to progress: Slow interconnects**

There is a second side effect of the miniaturization of transistors that threatens to stall its seemingly relentless progress. It's embarrassingly low-tech compared to dopants, p-n junctions, and digital logic.

It's the on-chip equivalent of wire. The metal interconnects between transistors cannot be reduced indefinitely in size without the risk of actually reducing switching speed.[7]

To be specific, as we note in Chapter 1, continuing to shrink gate sizes below about 0.065 micron (65 nm) is going to bring the interconnect problem to the forefront. At this point the most practical

---

[7] J.D. Meindl, "Beyond Moore's Law: The interconnect era," *Computing in Science and Engineering* 5 (January 2003), 20–24.

approach to further performance improvements has proven to be implementing single-chip multiprocessor architectures, discussed in depth in Chapter 2. This technique promises to let designers continue delivering advances in processing technology and architecture.

### MOSFET shrinks still pay

There are also architectural tricks that can be used in designing chips. Transistors of different dimensions can be used on the same chip, with only the fastest ones having the troublesome deep sub-micron dimensions. Having said all that, however, there are practical limits, primarily set by power dissipation, which have profound implications for future chip transistor densities and further increases in switching speeds.

However, these remarks should not be interpreted to mean that the historical trends are already at their end. More performance will be obtained by reducing MOSFET dimensions, not necessarily because the transistors are switching faster, but because their density increase provides more functions on a chip. The wireless handset industry will benefit from those trends as functions that previously requried several individual devices are consolidated into a single chip. The MOSFETs on these chips will be switching well below their GHz theoretical limits; hence power dissipation will be manageable.

# Appendix 1.2: Building multi-transistor logic gates

I N Chapter 1, we described the three basic logic gates used to build computers. Here we will illustrate a combination gate – a NAND built with MOSFETs. As shown in Figure A-1.2.1, it takes three transistors of the same n-channel type in this configuration.[1] A CMOS configuration for the same gate function requires four transistors.

The NAND is a combination of an AND gate and an inverter. The top MOSFET is connected to a voltage source of 10 volts which biases it in the *on* (conducting) state.

The truth table gives the logical combinations of this gate.

If both voltage A and voltage B are positive (1), then the bottom two transistors are turned *on* and hence conducting. This effectively connects the current flow from the voltage source through the top transistor to ground, and the output voltage V is 0. Hence it takes two positive (1) inputs to produce a (0) output.

On the other hand, if either of the two voltages at A or B is 0, then the current path is connected to the output, resulting in 10 volts at the terminal, which is defined as a "1."

[1] G. McWhorter, *Understanding digital electronics* (a Texas Instruments Learning Center Publication, 1978), pp. 2–17.

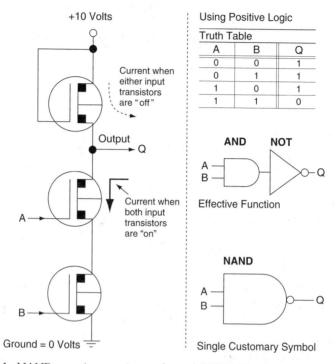

**A-1.2.1.** NAND gate input using n-channel MOSFETs. Courtesy of Texas Instruments (ref. 1).

# Appendix 1.3: MOSFETs in memory devices

AN important application of multi-transistor gates is the creation of building blocks that retain information, enabling data storage devices.[1]

The static random access memory (SRAM) requires six transistors per storage cell. The electrical configuration of the "unit" determines whether a bit is stored there or not. The MOSFET structures used in SRAMs are fairly standard and comparable to those used in building logic elements.

However, special structures are used to build the dynamic random access memory (DRAM), which stores a charge on individual transistors that incorporate a capacitor. This charge must be constantly refreshed in millisecond cycles as it leaks away.

At the other end of charge storage technology is FLASH (non-volatile) memory, which will maintain charge in the complete absence of a power supply. It requires special MOSFET structures very different from the standard devices.

These MOSFETs hold an electrical charge for a very long period of time in a special isolated capacitor.

Figure A-1.3.1 shows a highly simplified schematic of a special MOSFET p-channel transistor.[2] In contrast to a standard MOSFET, it has a silicon film imbedded within an insulating silicon dioxide film sitting on top of where the gate would normally be. This constitutes the "floating" gate of the p-channel MOSFET.

The basic operation by which the cell stores one bit is as follows.
- The device is programmed by injecting an electronic charge into the floating silicon gate (by applying a very short high-voltage pulse).

---

[1] B. Prince, *Semiconductor memories*, 2nd edition (New York: John Wiley & Sons, 1991).
[2] R. Zaks, "System components," *From chips to systems: An introduction to microprocessors* (Berkeley, CA: SYBEX, Inc., 1981), p. 139.

357

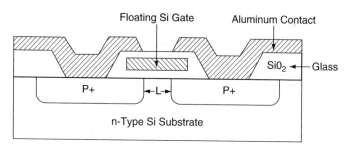

**A-1.3.1.** Schematic of basic MOSFET structure for non-volatile memory devices. Reprinted with permission from R. Zaks (ref. 2).

- This negative charge in the silicon film – the "floating gate" – induces a conducting p-type channel between the two p-type regions, turning it on.
- Hence, when this transistor is interrogated to determine whether a bit is stored there or not, its conducting state signals that a charge, a (1), is stored in that location.
- In the absence of a charge on the floating gate, the transistor is off, a (0), because no conducting channel is present between the two p-n junctions, and this is the result that is returned when the transistor is interrogated.

Practical non-volatile memories are considerably more complex. They require sophisticated circuits on the chip for programming, erasing the charge, and addressing.

A final note: Whatever the specific transistor architecture used to store bits, all memory devices rely on combinations of basic gates for their addressing, sensing, and communications tasks.

# Appendix 1.4: CMOS reduces logic gate power dissipation

D IGITAL processing needs lots of logic gates. MOSFET transistors, as noted in the main text, are the preferred way to create them. However, owing to excessive current leakage, early gates built with MOSFETs suffered from high power dissipation. Put millions of these devices on a chip and the constant current drain could cause overheating.

The CMOS architecture provided the solution to this problem, at the expense of more complex chip manufacturing, by dramatically reducing the gate power dissipation, enabling very large scale integrated circuits.

To demonstrate the principle behind CMOS, let's examine the differences between a simple inverter gate built without this architecture and one built with it.

We start with the *non*-CMOS inverter built with just one kind of MOSFET.[1] As Figure A-1.4.1 shows, two n-channel transistors are interconnected with the source of the bottom transistor connected to the drain of the top transistor. The top transistor is always on since its gate is always connected to the battery.

Look at the left side of the figure. When a value of 10 volts (the input voltage) is applied to the bottom transistor gate, a (1), it turns on, forming a conducting path from battery to ground. Therefore, the output voltage is zero, indicating a (0) state.

Now look at the right side of the figure. The input gate voltage is zero, the bottom transistor is off, and the output is connected to the battery through the top transistor. Hence the output voltage is 10 volts, a (1).

---

[1] G. McWhorter, *Understanding digital electronics* (a Texas Instruments Learning Center Publication, 1978), pp. 2–15.

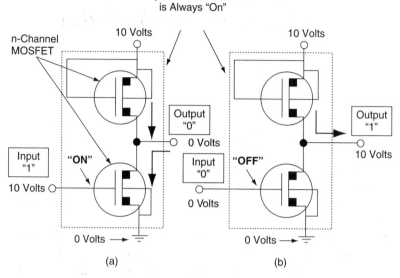

**A-1.4.1.** Inverter using n-channel MOSFETs. Courtesy of Texas Instruments (ref. 1).

This is simple enough, but it creates a huge problem when we want to put many of these gates on one chip in close proximity: constant power consumption. A small current is always flowing through the top transistor between the battery and the output, whether the lower transistor is on or off.

Since the top transistor never turns completely off, its wasted current leads to constant power dissipation. Put millions of these devices on a chip and the power dissipation can cause overheating. That limits the number of transistors on a chip.

Figure A-1.4.2 shows how the CMOS inverter works. We combine two different (complementary) MOSFETs, one n-channel (bottom) and the other p-channel (top). The gates of both transistors are connected to the input voltage.

Observe what happens in (a) when the positive input gate voltage is 10 volts applied to both transistors. The top p-channel transistor is *off* while the bottom n-channel transistor is *on*. As a result, the output voltage is zero because the conducting path is to ground through the turned-on bottom transistor.

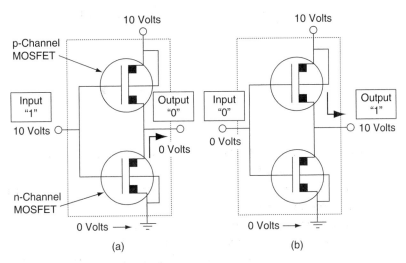

**A-1.4.2.** Inverter using CMOS structure.

Compare that to the right side of the figure. Now the input voltage is zero. The top transistor is *on*, while the bottom transistor is *off*. As a result, a conduction path is formed between the battery and the output where now 10 volts appear defined as a (1).

The only time a conduction path through the gate is formed is when a logic path has to be generated. Since the top transistor turns off when not needed, there is no significant leakage of current. In reality there is always some very small leakage current through transistors that are turned off, but it is orders of magnitude lower than in the non-CMOS inverter shown above.

In summary, the CMOS architecture greatly decreases the constant leakage current of transistors forming parts of gates on a chip as compared to earlier approaches. Less leakage translates into much lower power dissipation, and hence less heating of the chip.

The example above was chosen for simplicity. The same concept of using complementary transistors applies to all gate structures, but it does require that one or more transistors be added to form certain gates. CMOS made it possible to produce the microprocessors of today, densely populated with hundreds of millions of transistors.

As a result, CMOS structures are dominant in practically all integrated circuits, including processors, memory chips, and imagers.

# Appendix 1.5: Laser diode basics

THE fundamental physics of laser operation are similar for all media capable of laser operation, whether gases, insulators, or semiconductors. However, the operating details are vastly different in the various materials.[1]

The early lasers were made using special solid-state materials or gases confined within glass envelopes.[2] For example, important gas lasers use helium-neon or argon gas mixtures enclosed in a large glass tube 10 centimeters or more long. These are obviously bulky but useful as light sources when certain emission colors or high power are required – but not for the major applications that awaited semiconductor lasers.

The basic attraction of semiconductor lasers is their high atomic density, which makes possible laser operation in a volume which is *ten million times smaller than in a gas*. The idea of using semiconductors was discussed theoretically in the 1950s, but it took until 1962 for a laser diode device to be demonstrated.[3]

In semiconductors, lasing is possible only in certain classes of materials endowed with special natural properties – direct bandgap semiconductors. Lasing occurs as a result of a highly complex interactive process between very high densities of electrons and holes confined

---

[1] A comprehensive review is provided by G. P. Agrawal and N. K. Dutta, *Semiconductor lasers*, 2nd edition (New York: Van Nostrand Reinhold, 1993). B. E. A. Saleh and M. C, Teich (eds.), *Fundamental of Photonics* (New York: John Wiley & Sons, Inc., 1991).

[2] A history of early laser development can be found in J. L. Bromberg, *The laser in America: 1960–1970* (Cambridge, MA: The MIT Press, 1991).

[3] R. N. Hall, G. E. Fenner, J. D. Kingsley, T. J. Soltys, and R. O. Carlson, "Coherent light emission from GaAs junctions," *Physical Review Letters* 9 (1962), 366–368; M. I. Nathan, W. P. Dumke, G. Burns, F. H. Dill, Jr., and G. Lasher, "Stimulated emission of radiation from GaAs p- junction," *Applied Physics Letters* 1 (1962), 62–63; T. M. Quist, R. H. Rediker, RlJ. Keyes, W. E. Krag, B. Lax, A. L. McWhorter, and H. J. Ziegler, "Semiconductor master of GaAs," *Applied Physics Letters* 1 (1962), 91–92.

within a region where the intense radiation released by their recombination is also confined.

Laser emission has been obtained in semiconductor laser diodes in a spectral region ranging from the blue to the far-infrared. This makes semiconductors the materials of choice for most laser applications requiring intense light beams that can be modulated at high frequencies with high reliability and small size.

To produce a simple working device, the confined volume where lasing occurs needs to be bound in two sides by partially reflecting mirrors (a Fabry-Perot cavity) through which the laser emission is released to the outside. This is the structure shown in Figure 1.8. For applications requiring very stable control of the emitted spectrum, laser structures use internal gratings to form what are called "distributed feedback" lasers. We need not dwell on these architectural details. Our purpose is to understand the advances made possible by the introduction of heterojunction lasers which can meet the technological requirements for many applications. The most important fact is that many material combinations have been developed to produce heterojunction lasers covering a wide spectral emission range from the blue into the far-infrared.

The first semiconductor laser diode consisted of a special p-n junction structure in crystals of gallium arsenide (GaAs). Both sides of the junction used the same GaAs material. Called homojunction lasers, these devices operated best when refrigerated to low temperatures, but were impractical for room temperature operation owing to their excessive current requirements and poor reliability.

The introduction of heterojunction laser structures in the late 1960s enabled all of the current applications of these devices with a wide range of materials. However, the fundamental principles of operation can be understood from a very simple double-heterojunction structure based on GaAs and higher bandgap energy aluminium gallium arsenide (AlGaAs).

Figure A-1.5.1 shows two highly-simplified laser diode structures to illustrate the difference between the original homojunction and a heterojunction structure. Part (a) shows a simple p-n junction laser, where the laser action occurs in a volume that is of relatively uncontrolled thickness d. This is approximately 2 microns in the example shown.

Part (b) shows, for illustration purposes, a simple double-heterojuncton laser. The p-n junction has been replaced by a p-n heterojunction of

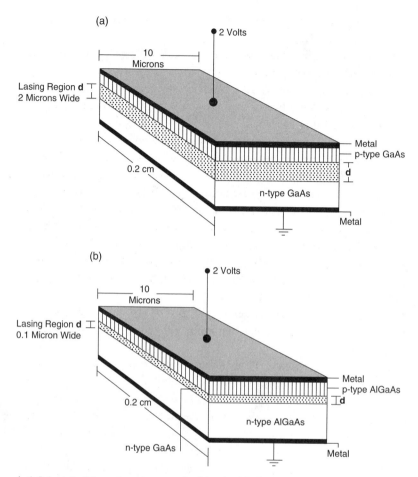

**A-1.5.1.** (a) Homojunction and (b) double-heterojunction laser diode structures showing the recombination region layer of thickness d.

AlGaAs/GaAs, and an additional n-n heterojunction of GaAs/AlGaAs has been placed within 0.1 micron of the first.

One result of this approach is that the width of the active region in which laser action occurs is reduced from about two microns to approximately 0.1 micron. Furthermore, an internal refractive index profile is formed for guiding and confining the radiation produced as shown in Figure A-1.5.2. There are also other factors which allow more of the internally-generated light to escape, as internal absorption of light

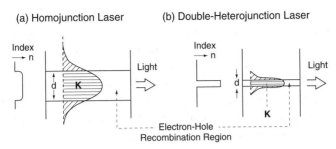

**A-1.5.2.** Schematic cross-section of a double-heterojunction laser showing the refractive index distribution perpendicular to the junction plane and intensity distribution of the fundamental transverse mode with a fraction K of the radiant energy within the recombination region of width d.

is reduced by confinement of the radiation as it propagates toward the edge of the device. As an approximation, because of the much improved carrier and radiation confinement resulting from the thin lasing region, as much as a twenty-fold reduction in the laser volume is obtained. This translates into a corresponding reduction in the diode current flow needed to sustain lasing.

To better understand the impact of the heterojunctions we need to look at the internal distribution of the radiation as it travels in the device. Figure A-1.5.2 shows the refractive index profile inside the homojunction and double heterojunction laser to illustrate the important difference between them.

In (a) we see the internal refractive index profile of the homojunction lasers where the electron-hole recombination occurs in the region d, but the radiation produced is poorly confined.

In (b) we see the effect of confining the lasing region between the two heterojunctions. Not only is the recombination region very thin because the electron-hole recombination is confined by two heterojunctions, but in addition the sharply greater difference in refractive index at the boundaries focuses the radiation and produces much lower losses as the radiation propagates toward the edge of the laser where it is emitted. The radiation spreads outside that region to some extent, but in a very controlled way.

The end result is that the current density that this structure requires for sustained laser emission is drastically reduced, by a factor that is nearly (but not quite) proportional to the reduced laser volume.

Figure 1.7 shows the historical reductions in the threshold current density as the technology continued to be refined, reaching operating currents so low that they proved to be no hindrance to the most important commercial or consumer laser applications.

For modern optical communications, heterojunction lasers require emission in the 1.55 micron range and are constructed from combinations of indium, gallium, arsenic, and phosphorous compounds.

As new materials get developed, such as lasers using gallium nitride compounds operating in the blue spectral region, new materials challenges have to be overcome to reach acceptable operating current levels. It usually takes many years to master the manufacturing technologies for new laser materials.

# Appendix 1.6: Light-emitting diodes (LEDs)

S EMICONDUCTOR light emitting devices are grouped into two categories:

- Light emitting diodes (LEDs), which are basically general-purpose replacements for vacuum tube-based light emitters.
- Laser diodes that emit sharply focused monochromatic light beams.

We have already discussed laser diodes. Here we will look briefly at LEDs, which have come to dominate important segments of the lighting industry.

Light emitting diodes consist of p-n junctions in a special class of semiconductors that are capable of emitting light of many colors ranging from the infrared into the violet. Furthermore, of great practical importance is the fact that by combining LEDs of red, blue, and green, or using special phosphor coverings, white light can be produced, with conversion efficiencies from electrical input to useful light output that are very competitive with conventional vacuum-based devices.

When we discussed p-n junctions, we noted that one of their most valuable properties is their ability to inject carriers across the p-n interface. This is the principle behind LEDs. When forward-biased, they emit various colors of light, ranging from the blue into the infrared, depending on the semiconductor materials used.

Semiconductors useful as LEDs and lasers have special structures. As noted earlier when we discussed lasers, they are direct bandgap materials. Within them the recombination of an injected electron with a hole (in effect the capture of the free electron by an empty spot in the atomic outer orbit) results in the release of energy in the form of a photon. A photon is a unit of light energy. The value of this photon, hence the color emitted, depends on the bandgap energy of the material. For example, in gallium arsenide with a bandgap energy of about 1.4 electron-volts, the emitted light is in the near-infrared, about 0.9 micron wavelength.

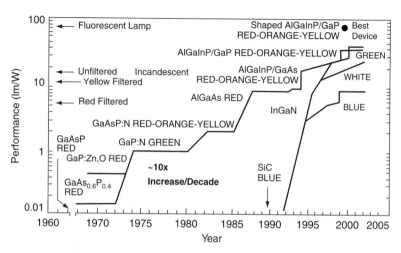

**A-1.6.1.** Improvement in visible LED efficiency since the 1960s. LEDs have averaged efficiency improvements of ~10x/decade over this time frame. Today's devices are significantly more efficient than unfiltered incandescents with the best devices approaching fluorescent lamp efficiencies. Reprinted with permission from *physica status solidi (a)* and F. Steranka. Copyright © 2002 Wiley-VCH Verlag GmbH & Co. KG (ref. 1).

The most important semiconductor crystals having the right optical properties are alloys containing elements from the third and fifth column of the periodic table of elements. These include gallium arsenide, gallium arsenide phosphide, gallium nitride, and others.

As versatile as it is in other applications, silicon is not one of the semiconductors used for light emitting diodes. It simply has the wrong crystalline structure for this effect. In silicon the recombination of a free electron with a hole results in the release of small amounts of heat, not a photon.

Since the realization of the first practical red-emitting LEDs in the 1960s, enormous improvements have been made in the devices, covering a very broad range of emitted colors. Figure A-1.6.1 shows the evolution of the efficiency and color availability of LEDs since the 1960s.[1]

---

[1] F. M. Steranka, J. Bhat, D. Collins, L. Cook, M. G. Craford, R. Fletcher, N. Gardner, P. Grillot, W. Goetz, M. Keuper, R. Khare, A. Kim, M. Krames, G. Harbers, M. Ludowise, P. S. Martin, M. Subramanya, T. Trottier, and J. J. Wierer, "High power LEDs – Technology status and market applications," *physica status solidi (a)* 194 (2002), 380–388.

Some devices are exceeding the luminous efficiency of vacuum-based light emitters such as incandescent or even fluorescent bulbs. In fact, the quality and cost of LEDs have reached the point where they can selectively replace some vacuum tube-based light sources in applications such as automobile headlights and traffic signals.

Recently several TV makers have announced rear-projection sets that replace the conventional projector bulb with an LED array for longer life, better power, and quiet operation without a cooling fan.

# *Appendix 1.7: Photodetectors*

S EMICONDUCTOR devices for sensing light levels are in wide use, and their properties are tailored to specific applications. They range in complexity from the simplest light sensors, used in door openers, to sophisticated devices that detect ultra-short, very low-level communications signals that have traveled hundreds of miles over an optical fiber.

As mentioned in the main text, light sensors are also the core sensing element in imaging devices, where millions of sensors are built into a single chip. Finally, solar energy converters (solar cells) are in the same class of device.

We learned in Chapter 2 that a photodetector is usually a reverse-biased p-n junction. The type of semiconductor material used for the junction is determined by its intended application, particularly with regard to the wavelength of the light to be detected and the response speed desired.

The operational principle is that a photon of incident light absorbed in the vicinity of the p-n junction, and within the depletion zone, generates free carriers that give rise to a current proportional to the intensity of the incident light. Figure A-1.7.1 shows the p-n junction detector in the dark and with incident light. The current through the device increases with light intensity.

Silicon photodetectors are widely used to capture visible light, whereas III-V compound devices are used to detect signals in the infra-red portion of the spectrum. Most detectors used in fiber optic communications systems, where the laser emission is in the 1.5 micron (near-infrared) spectral region of low fiber absorption, are made of indium gallium arsenide.

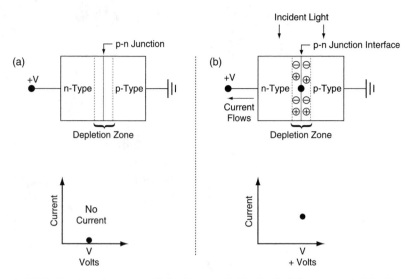

**A-1.7.1.** P-n junction used as light detector. (a) In the dark (no current); (b) with incident light which results in current flow proportional to light intensity.

# Appendix 1.8: Making fiber optic cables

To gain an appreciation of how one builds optical fibers for different communications applications, look at Figure A-1.8.1, which illustrates three different types of fibers.[1]
Each of these types is distinguished by its refractive index profile. The refractive index profile is a key part of fiber design because it determines the product's ability to propagate short pulses and maintain their shapes over long distances.

The topmost illustration shows the simplest stepped-index profile fiber (also called multimode). This is the least costly type to manufacture, but it has the poorest pulse propagation characteristics. It is used for transmission over short distances.

Single mode fiber with a tailored stepped refractive index profile, shown in the second illustration, is much better at keeping the pulse shape intact, but it is costlier. This type of fiber is used for high-data-rate communications over long distances.

Finally, in the illustration at the bottom of the figure, we have a multimode, graded-index fiber that falls between the two profiles above it. This fiber has applications in intermediate distance transmission.

In all three examples the glass fibers are encased in claddings. Large numbers of individual cladded fibers are enclosed in cables that protect them from external forces and the elements. Special techniques exist for creating cables that can survive undersea deployment and other demanding environments.

---

[1] A. G. Chynoweth, "The fiber lightguide," *Physics Today* (May 1976), 28–37.

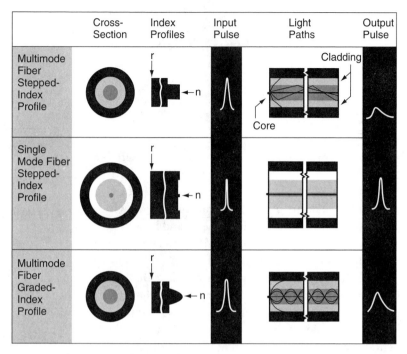

**A-1.8.1.** Three types of optical fibers. (a) Multimode fiber; (b) single-mode fiber; and (c) multimode graded-index fiber. Reprinted with permission from *Physics Today*. Copyright © 1976 American Institute of Physics (ref. 1).

# Appendix 1.9: Principles of LCD displays

FLAT-SCREEN displays based on organic materials dominate the display device field. Chief among them for its commercial importance is the liquid crystal display, or LCD.

Liquid crystals are a class of organic materials with very interesting properties. Their molecular alignment (and hence their optical properties) change reversibly with temperature, as well as under the effect of moderate electric or magnetic fields.[1]

At high temperatures these materials exhibit the properties of liquids, while below a critical temperature they assume more of the properties of solids. For example, as a result of this transformation, a liquid crystal film can change reversibly from a milky fluid, opaque to light, to one which is transparent to visible light.

The discovery of liquid crystals dates to the nineteenth century. However, the materials languished without serious applications until the late 1960s, when research at RCA Laboratories found an application for them in flat-screen displays. The researchers achieved this by applying electric fields to thin liquid crystal films to change the optical properties of the materials.[2]

This discovery made two types of display applications possible as materials were developed. First, applying an electrical field to certain classes of liquid crystals changes them from light absorbers to good reflectors. Second, in other materials, the applied field can make the film either opaque or transparent to visible light; hence the film can become a "light valve."

---

[1] For a complete description see G. W. Gray, *Molecular structure and properties of liquid crystals* (London: Academic Press, 1962).

[2] G. H. Heilmeier, L. A. Zanoni, and L. A. Barton, "Dynamic scattering: A new electro-optic effect in certain classes of nematic liquid crystals," *Proceedings of the IEEE* 56 (1968), 1162–1170.

Flat-screen display technology has proven to be crucial to the mass acceptance of sophisticated electronic systems. There would be no personal computers, digital cameras, cellular handsets, or dozens of other portable products without the ability to display high-resolution images on inexpensive, lightweight displays. Of course, transistors play a key role in enabling LCD displays as part of their data stream and addressing management.

The simplest structure for making liquid crystal displays consists of sandwiching a thin film of the crystal material between two glass plates, each side of which is covered with a transparent conductor. To form an imager, we divide the "sandwich" into a grid of individual pixels, each of which can be a recipient of the voltage necessary to change the properties of the material within its borders.

For example, we can choose to make an array of two hundred by three hundred pixels. Each pixel can be accessed by a transistor circuit that applies a temporary voltage to that pixel. Assuming that there is uniform light incident on the device, images will be created by the pattern of pixels that do or do not reflect the light to our eyes, based on whether they have been subjected to the applied electric field. Color filters can be incorporated to turn the image from monochrome into full color.

If we are interested in creating images *without uniform incident light*, we can use liquid crystals that shift from transparent to opaque with the application of an electric field. This type of display requires that a light source, such as LEDs for small displays or fluorescent devices for larger ones, be placed behind the screen, so it can shine through the transparent pixels to create the image.

What we have described is really primitive. Today's versions are far more sophisticated. Modern "active matrix" displays are built with thin films of silicon deposited on glass and containing transistors arrayed in complex integrated circuits. In effect the thin film electronic circuitry is built as part of LCD display.[3]

Manufacturing such displays is very demanding and involves technologies as complex as those needed to produce silicon integrated circuits. Production plants cost billions of dollars.

---

[3] A useful text on modern LCDs is W. den Boer, *Active matrix liquid crystal displays* (Amsterdam: Elsevier, 2005).

LCDs face competition from new flat-panel technologies. These include displays using organic light emitting diodes (OLEDs) and, for large TV displays, vacuum plasma devices built using huge arrays of tiny electrodes that are image-energized to produce small streams of electrons that energize phosphors in order to produce the color effect. Such devices fall outside of our consideration of solid-state imagers and do not impact the mass market for LCD flat-panel displays.

# Appendix 2.1: The demise of analog computers

I n Chapter 1 we pointed out the difficulty of using vacuum tubes for large-scale digital data processing.

This observation, while perfectly true, begs the question, why shift to digital in the first place? We live in the natural world, which is defiantly analog. It seems logical that analog systems would process information from the analog world more efficiently than their digital counterparts.

On the surface they do seem more efficient. In analog electronic systems a smoothly variable electrical quantity (such as voltage, capacitance, or current) is used to represent information. The value of the electrical property is directly proportional to the magnitude of the physical signal being processed or transmitted.

Suppose, for example, you want to measure the intensity of light in a room. You place a photodetector (an analog light sensor) there. When light hits the device, it responds either by generating an electrical current or by permitting such a current to flow. As the light gets stronger, the current flowing in the system becomes greater as well.

The current increases in a linear fashion, allowing you to track and record the level of light falling on the photodetector. The term "linear" is used to describe analog systems because, under ideal conditions, their electrical information varies in line with the received signal.

Human beings are analog systems, too. We respond in a linear fashion to external analog signals such as temperature, sound, light, and images. Our brains are analog computers that interpret and analyze these signals, then generate output signals, which are delivered through our nervous systems.

Since our brains are analog, the question of why go digital applies equally well to computing systems. Why not simply replicate the human brain in an analog computer system?

In fact, the first practical computers were analog. Well before digital computers were widely available, there were special-purpose analog units that computed artillery trajectories and solved complex scientific equations in heat flow and aerodynamics. These computers operated by manipulating voltage levels in circuits that combined, at various times, vacuum tubes, transistors, diodes, amplifiers, resistors, and capacitors.

By today's standards analog computers were quite limited. When more flexible and powerful digital processors became available, stand-alone analog machines simply disappeared.[1] A comparison between digital and analog will make clear the reasons for this sudden shift.

## The digital advantage

We now take the superiority of digital computers for granted, but we rarely consider why they vanquished analog processors. Here are some of the reasons.

### *Scalability*

As we saw in Chapter 1, only three basic logic gates are needed to perform any logic or arithmetic function using Boolean algebra. Each CMOS gate requires, on average, four transistors, as shown in Appendix 1.4.

As a result, large computations are *conceptually* simple. They can be carried out through multiple arrays of interconnected gates. To increase the processing power of a digital computer, you just increase the number of available gates.

In practice, of course, matters get more complicated, and the architecture of the device becomes all-important. But in theory, as long as you can manage the interconnections that carry data traffic, you can continue to increase processing power. There are no theoretical limits to the size or power of a digital computer.

Analog computers, on the other hand, must deal with complex interrelated changes in signal levels. As the computer gets larger, it becomes less reliable, since its greater complexity increases the chances that the analog signals will interfere with each other. This makes scaling enormously complicated and very costly.

---

[1] While stand-alone analog computers may be history, analog computation is very much alive as part of some digital integrated circuits. Among other functions it is used for power management and for analyzing wireless signals. If done in the digital domain, these processes would require large, complex digital signal processing circuitry on the ICs. Keeping the signals in analog yields superior results and smaller chips.

*Flexibility*

The function of an analog computer is usually determined by its hardware. By contrast, a digital computer can be adapted to different functions through the use of programs stored in software or firmware. You don't have to change hardware when you change tasks.

*Data integrity and processing precision*

Internal and external noise and interference can disrupt the linear nature of analog information, especially when the signals carrying that information are weak. Digital processors, on the other hand, are isolated from analog variations by their very nature, and therefore not subject to interference.

An analog processor also has difficulty maintaining precision when the values are very small. The mathematical basis of digital processors allows them to handle very small and very large values with equal accuracy.

*Data storage*

Digital data (input and output) can be stored indefinitely in various types of memory devices. Analog computers store information only for relatively short periods of time.

*Communications reliability*

Digital communications have overwhelming advantages over analog signal transport. A digital signal can be transmitted over any distance, provided its bits can be individually identified at the receiving end.

In an analog communications system, by contrast, irreversible changes in the waveform cause the signal to deteriorate over distance. The signal is also more likely to encounter interference from the analog world around it. In either case information is lost.

The modern communications infrastructure could never have been built with analog systems.

**Achieving digital accuracy**

Digital signal and data processing is done in binary format. Therefore digital systems must convert analog signals into ones and zeros so they can be processed in the digital domain.

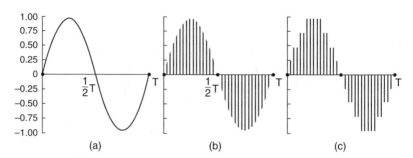

**A-2.1.1.** (a) A sine wave; (b) sampling the sine wave; (c) quantizing the samples to 3 bits which produces eight distinct values for each sample. Copyright © 1996 Pearson Education, Inc. Reprinted by permission of Pearson Education, Inc., Upper Saddle River, NJ (ref. 2).

While this seems like an extra step, it permits easier, more accurate processing, provided the conversion is done with high enough precision. When an analog signal is digitized, the precision of the digital signal increases with its sampling rate and the number of bits used to describe the values it represents.

Take the simple example of digitizing a sine wave (Figure A-2.1.1). The wave is shown in (a) and the digitizing process in (b). If we digitize the signal using 3-bit numbers, we can only obtain eight values, from $-1.00$ to $+1.00$, in steps of $0.25$ (c).

We can get much higher precision by adding bits. For example, with 8 bits we can get 256 values for each sample. With 16 bits we get 65,536 values for each sample.[2] It would be difficult to achieve this level of precision in an analog system.

Precision is not the only advantage digital systems have over their analog equivalents. Digitized data can also be processed repeatedly without having values shift or deteriorate. After all processing is done, digital signals can be converted back to analog if the results require human interface.

In short, digital systems offer the best of both worlds: the precision, flexibility, and reliability of a system based on mathematics, and the ability to interact with our analog world.

---

[2] Figure and calculations from A. S. Tanenbaum, *Computer networks*, 3rd edition (Upper Saddle River, NJ: Prentice Hall PTR, 1996), p. 725.

# Appendix 2.2: IP, TCP, and the Internet

W E already know that special computers called routers direct data traffic on a packet-switched network, and that computers communicate through special, industry-defined machine languages (protocols). Here we examine the protocols that underlie packet networks in general and the Internet in particular.

The protocol that has reached star status for communications systems is the Internet Protocol (IP), which grew out of the early ARPANET work. Networks designed to operate with the IP protocol use a 32-bit address embedded in each packet header to move it from source to destination.

Every computer on the Internet has an IP address. That includes each router. Since routers transmit digital streams of bits in the form of packets, the physical implementation of the IP protocol involves coding headers attached to the packets. These headers provide the address of the sending and receiving computers.

The packets travel a path from one router to another on the way to their destination. The routers direct this trip, using stored lookup tables to find paths to the destination IP address in the packet headers.

But even IP doesn't work alone. To provide reliable internetworking on the Internet, it is closely linked with TCP (Transmission Control Protocol). Let us look at how IP and TCP work together.

While the router lookup tables are continuously updated, that doesn't prevent packets from getting lost or delayed. If a target router is temporarily overloaded or down, the packets sent through it could lose their way, and the network would be unreliable.

To prevent this from happening, TCP provides the equivalent of certified mail. The instructions imbedded in the TCP part of the headers ensure that packets are handled appropriately. In effect, it establishes a "virtual" connection between the source of the data and its destination.

TCP also supplies an acknowledgement that the packet has arrived at the destination, and that it has not been corrupted during its passage. If a packet is lost or damaged, a makeup packet or replacement will be sent.

The maximum content of each packet, including its header, is limited to 65,535 bytes (8 bits in each byte). The TCP portion of the header contains much more information about source, destination, and correct packet sequence than the IP portion. It is the main reason that the packets can be reassembled at the destination in the proper order.

Another aspect of the TCP protocol is the ability it provides to encode headers with information specifying levels of security, and to manage the parallel flow of related packet streams.

IP may get star billing, but TCP does a lot of the behind-the-scenes work.

Protocols designed to label packets for special treatment continue to be developed to further provide a quality of service beyond that of TCP/IP. For example, a standard has been established called multi-protocol label switching (MPLS).

MPLS ensures that routers provide appropriate treatment to related packets. It adds a 32-bit header to the packet that contains prioritizing information and the type of content being transported. In effect, packets having the appropriate label headers get priority routing as they navigate through the networks.

As a result, therefore, IP-enabled networks are fully capable of reliably transporting any kind of data. This is a transformational achievement, allowing the connectionless network model to provide the same functionality as the PSTN, but with far more flexibility. This changes the game for the communications industry.

# Appendix 2.3: Building
## an object-oriented program

To get a feel for object-oriented architectures, consider a very simple program designed to update the contents of a particular master file. Figure A-2.3.1 shows the "old" way of architecting such a program, by logical decomposition of the process into steps that are sequentially linked to produce the program.[1] To produce this program, each element of the process must be coded in tightly-linked sequences.

By contrast, in the object-oriented architecture shown in Figure A-2.3.2, the program is divided into abstractions of the key elements.[2] Instead of steps, the problem is now defined in terms of autonomous interconnected objects. Each object has its real-world mission defined, and is given its operational orders and required variables through the bus.

Program development using object-oriented technology can be divided into four interdependent elements.

- The first critical task is the definition of program requirements. This requires close communication with the intended users and an understanding of their applications.
- Program architecture is developed on the basis of these requirements. The skill of the architect determines the ultimate quality of the program and the time required to complete it.
- The production of the program proceeds, with coding by programmers who write within tight guidelines provided by the program architects.
- Finally, the completed program must be tested. Note that testing can account for between 40 and 50 percent of the cost of developing a large program.

---

[1] G. Booch, *Object-oriented design with applications* (Redwood City, CA: The Benjamin/Cummings Publishing Company, Inc., 1991), p. 15.

[2] Booch, *Object-oriented design*, p. 16.

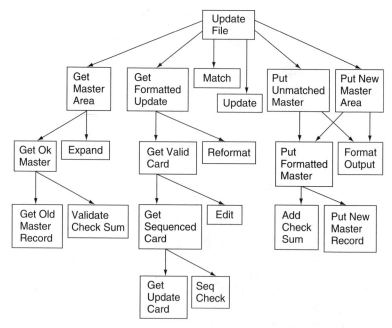

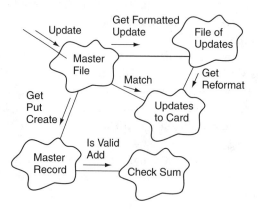

Program definition and architecture require the most skilled and experienced people. These two steps produce virtually all of the program's intellectual property value. Hence, these functions are very rarely outsourced.

Code writing and program testing require the most manpower, but a lower level of detailed expertise. This division of labor has made possible the partial outsourcing of software development to lower-cost geographies like India, where software "factories" employing thousands of people have been set up in the past few years to write and test software code.

In effect, software production is becoming analogous to the equipment business. The concept, product design, and architecture are done within an enterprise, while the more mechanical processes of assembly and final product test are frequently in the hands of outside contractors.

# Index